T0299230

Sustainable Manufacturing

Sustainable manufacturing is a key component in the engineering industry, helping to decrease emissions, costs, and energy use. Through examining how to successfully implement sustainability within Industry 4.0, *Sustainable Manufacturing: An Emergence in Industry 4.0* covers recent innovations in topics, including circular economy, supply chains, waste elimination, and recycling.

This edited collection is a cutting-edge assessment of the barriers preventing the implementation of sustainable manufacturing in industry. Highlighting basic definitions and terminologies within sustainability and manufacturing, this book covers topics that include interactive design, remanufacturing, cleaner production, and optimization. It also features modern technologies currently revolutionizing the industry, such as robotics and 3D printing. Using case studies to illustrate success stories in which products have been created using sustainable processes, this book also includes technical notes and experimental results from a wide variety of international contributors.

This book is relevant to anyone working in the mechanical engineering, manufacturing and industrial engineering, and materials science industries.

Manufacturing Design and Technology

Series Editor

J. Paulo Davim

This series will publish high quality references and advanced textbooks in the broad area of manufacturing design and technology, with a special focus on sustainability in manufacturing. Books in the series should find a balance between academic research and industrial application. This series targets academics and practicing engineers working on topics in materials science, mechanical engineering, industrial engineering, systems engineering, and environmental engineering as related to manufacturing systems, as well as professions in manufacturing design.

Non-Conventional Hybrid Machining Processes: Theory and Practice
Edited by Rupinder Singh, J. Paulo Davim

Functional Materials and Advanced Manufacturing: 3-Volume Set
Edited by Chander Prakash, Sunpreet Singh, J. Paulo Davim

Industry 4.0: Challenges, Trends, and Solutions in Management and Engineering
Edited by Carolina Machado, J. Paulo Davim

Sustainable Manufacturing for Industry 4.0: An Augmented Approach
Edited by K. Jayakrishna, Vimal K.E.K., S. Aravind Raj, Asela K. Kulatunga, M.T.H. Sultan, J. Paulo Davim

Industrial Tribology: Sustainable Machinery and Industry 4.0
Edited by Jitendra Kumar Katiyar, Alessandro Ruggiero, T V V L N Rao, J. Paulo Davim

Innovative Development in Micromanufacturing Processes
Edited by Pawan Kumar Rakesh and J. Paulo Davim

Tribology in Sustainable Manufacturing
Edited by Jitendra Kumar Katiyar, TVVLN Rao, Ahmad Majdi Abdul Rani, Mohd Hafis Sulaiman, J.P. Davim

Sustainable Manufacturing: An Emergence in Industry 4.0
Edited by Kamalpreet Sandhu, Sunpreet Singh, Ranvijay Kumar, J. Paulo Davim, Seeram Ramakrishna

For more information about this series, please visit: https://www.routledge.com/Manufacturing-Design-and-Technology/book-series/CRCMANDESTEC

Sustainable Manufacturing

An Emergence in Industry 4.0

Edited by
Kamalpreet Sandhu
Sunpreet Singh
Ranvijay Kumar
J. Paulo Davim
Seeram Ramakrishna

Designed cover image: www.shutterstock.com

First edition published 2025
by CRC Press
2385 NW Executive Center Drive, Suite 320, Boca Raton FL 33431

and by CRC Press
4 Park Square, Milton Park, Abingdon, Oxon, OX14 4RN

CRC Press is an imprint of Taylor & Francis Group, LLC

ISBN: 978-1-032-31309-2 (hbk)
ISBN: 978-1-032-31313-9 (pbk)
ISBN: 978-1-003-30912-3 (ebk)

DOI: 10.1201/9781003309123

Typeset in Times
by SPi Technologies India Pvt Ltd (Straive)

Contents

Preface

Sustainable Manufacturing: An Emergence in Industry 4.0 aims to present various outbreaks of novel concepts and methodologies dealing with sustainable manufacturing in Industry 4.0. This book is compiled to provide a comprehensive knowledge review of the key fundamentals, recent innovations, tools, ideologies, and research methodologies covering the diversified implications of sustainability concepts in modern industries. This book also includes today's popular sustainability concepts, which cover zero-waste manufacturing, life cycle assessment, circular economy, etc.

Editors

Kamalpreet Sandhu is currently working as an Assistant Professor/Head of Design Labs in the Product and Industrial Design Department at Lovely Professional University (LPU), Phagwara, Punjab, India. He edited various books, followed by *Revolutions in Product Design for Healthcare* (Springer, 2022), *Emerging Application of 3D Printing during COVID-19* (Springer, 2022), *Application of 3D Printing in Biomedical Engineering* (Springer 2021), and *3D Printing in Podiatric Medicine* (Elesvier, 2023). He has contributed extensively to human factors, product design, entrepreneurship, and 3D printing, with publications appearing in the Defence Life Science Journal; the International Journal of Human Factors and Ergonomics; the Journal of Manufacturing Process, Materials, Material Engineering, and Performance; and International Journal of Interactive Design and Manufacturing. He serves on the editorial review board of the International Journal of Technology and Human Interaction, The Design, and Frontiers in Manufacturing Technology. He is also a lifetime member of the Indian Society of Ergonomics.

Sunpreet Singh is a researcher at the National University of Singapore's Department of Mechanical Engineering. His research focuses on additive manufacturing and the application of 3D printing in developing biomaterials for clinical use. Singh has authored over 150 research papers and 27 book chapters, with publications in prestigious journals, such as the Journal of Manufacturing Processes and Composite Part: B. He is also the editor of three books on bio-manufacturing and 3D printing in biomedical engineering. Collaborating with esteemed institutions and experts, Singh contributes significantly to the field. He is a Guest Editor for various journals and actively participates in advancing the field of metrology in materials and advanced manufacturing.

Ranvijay Kumar is an Assistant Professor at the University Centre for Research and Development, Chandigarh University. He has received PhD in mechanical engineering from Punjabi University, Patiala. Additive manufacturing, shape memory polymers, smart materials, friction-based welding techniques, advanced materials processing, polymer matrix composite preparations, reinforced polymer composites for 3D printing, plastic solid waste management, thermosetting recycling, and destructive testing of materials are the skills of Dr. Kumar. He has won the prestigious CII MILCA award in 2020. He has co-authored more than 60 research papers in science citation-indexed journals, and 48 book chapters and has presented 20 research papers in various national/international level conferences. He has contributed extensively to additive manufacturing literature with publications appearing in reputed journals. He has edited six books published with leading publishers. In the years 2021 and 2022, Dr. Kumar was listed in the world's top 2% of scientists by Stanford University.

J. Paulo Davim received his Ph.D. degree in mechanical engineering from the University of Porto in 1997 and aggregation from the University of Coimbra in 2005. Between 1986 and 1996, he was a Lecturer at the University of Porto. Currently, he is an Aggregate Professor at the Department of Mechanical Engineering of the University of Aveiro and the Head of MACTRIB – Machining and Tribology Research Group. He has more than 24 years of teaching and research experience in manufacturing, materials, and mechanical engineering. He is the Editor in Chief of four international journals, Guest Editor, Editorial Board Member, Reviewer, and Scientific Advisor for many international journals and conferences. In addition, he has also published more than 250 articles in ISI journals (150 with an h-index of 14) and conferences.

Seeram Ramakrishna is a world-renowned professor and scholar of cross-field at the National University of Singapore (NUS). He made seminal contributions in understanding and enhancing the biological, chemical, electrical, electronic, mechanical, and physical responses of nanofibers | nanomaterials. He is among the World's Most Influential Minds (Thomson Reuters) and Clarivate recognized him among the Highly Cited Researchers since 2014. His publications to date received an h-index of 196 and 186,213 citations. Highest professional distinctions include an elected Fellow/Academician of the Royal Academy of Engineering (FREng), UK; the Singapore Academy of Engineering; the Indian National Academy of Engineering, India; the Chinese Academy of Engineering, China; and the ASEAN Academy of Engineering & Technology. He is an elected Fellow of AAAS, ASMInternational, ASME, AIMBE, USA; IMechE and IoM3, UK; ISTE, India; and IUBSE (FBSE). He received a PhD from the University of Cambridge, UK, and TGMP from Harvard University, USA. He received advanced research experiences from MIT and Johns Hopkins University, USA and KIT, Japan.

Contributors

A. A. Adeleke
Department of Mechanical Engineering
Nile University of Nigeria
Abuja, Nigeria

Ketan Badogu
Department of Mechanical Engineering
Chandigarh University
Mohali, India

Abhishek Bangre
Lovely School of Architecture and Design
Lovely Professional University
India

Sagarika Bhattacharjee
Department of Metallurgical and Materials Engineering
Indian Institute of Technology Ropar
Rupnagar, India

Harpreet Kaur Channi
Department of Electrical Engineering
Chandigarh University
Mohali, India

Jasgurpreet Singh Chohan
Department of Mechanical Engineering
Chandigarh University
Mohali, India, 140413
and
University Center for Research and Development
Chandigarh University
Mohali, India

P. P. Ikubanni
Department of Mechanical Engineering
Landmark University
Omu-Aran, Nigeria
and
Landmark University, SDG 9, Industry, Innovation, and Infrastructure Research Cluster
Omu-Aran, Kwara State, Nigeria

Kanwaljit Singh Khas
Lovely School of Architecture and Design
Lovely Professional University
India

Khushwant Kour
Department of Mechanical Engineering
Chandigarh University
Mohali, India

Rasleen Kour
Research Scholar, Department of Humanities and Social Sciences
Indian Institute of Technology Ropar
Rupnagar, India

Pulkit Kumar
Department of Electrical Engineering
Chandigarh University
Mohali, India

Ranvijay Kumar
Department of Mechanical Engineering
Chandigarh University
Mohali, India
and
University Center for Research and Development
Chandigarh University
Mohali, India

Vinay Kumar
Department of Mechanical Engineering
University Centre of Research and Development
Chandigarh University
Mohali, India

M. Malathi
Metal Extraction and Recycling Division
National Metallurgical Laboratory
Jamshedpur, India

H. O. Muraina
Department of Materials and Metallurgical Engineering
University of Ilorin
Ilorin, Nigeria

J. K. Odusote
Department of Materials and Metallurgical Engineering
University of Ilorin
Ilorin, Nigeria

M. Oki
Greenfield Creations Ltd.
Benue Close, Agbara, Ogun State, Nigeria

D. Paswan
Metal Extraction and Recycling Division
National Metallurgical Laboratory
Jamshedpur, India

Nishant Ranjan
Department of Mechanical Engineering and University Centre for Research and Development
Chandigarh University
India

Kamalpreet Sandhu
Department of Product and Industrial Design
Lovely Professional University
Phagwara, India

Ankan Shrivastava
Department of Mechanical Engineering
Chandigarh University
Mohali, India

Harmanpreet Singh
Department of Mechanical Engineering
Indian Institute of Technology Ropar
Rupnagar, India

Rupinder Singh
Department of Mechanical Engineering
Chandigarh University
Mohali, India

Rupesh Surwade
Lovely School of Architecture and Design
Lovely Professional University
India

Vishal Thakur
Department of Mechanical Engineering
Chandigarh University
Mohali, India

1 Zero-Waste Manufacturing and Cleaner Production

Sustainable Approach by Additive Manufacturing

Nishant Ranjan and Vinay Kumar
Chandigarh University, Mohali, India

1.1 INTRODUCTION

Conventional/traditional manufacturing processes such as casting, molding, and subtractive manufacturing often lead to significant emissions, energy consumption, and material waste, resulting in resource depletion and environmental degradation (Fraţila and Rotaru, 2017; Hegab et al., 2023). The idea of zero-waste manufacturing and cleaner production has attracted more attention as a sustainable manufacturing process to solve these problems. In contrast to cleaner production, which focuses on reducing the environmental impact of production through better use of resources and reduced pollution, zero-waste manufacturing seeks to minimize or eliminate waste throughout the manufacturing process, from the extraction of raw materials to the disposal of finished products (Singh et al., 2017; Kurniawan et al., 2021; Pietzsch et al., 2017). For reducing waste generation and environmental impact while utilizing resources, cleaner production, and zero-waste manufacturing are developing ideas in the area of sustainable manufacturing (Gupta et al., 2016; Sanchez et al., 2020). When compared to conventional manufacturing processes, additive manufacturing (AM) allows the fabrication of complex and difficult shapes and sizes easily (Javaid et al., 2021; Huang et al., 2013; Ngo et al., 2018).

There has been increasing research interest in investigating the potential of AM for sustainable manufacturing, with a focus on zero-waste manufacturing and cleaner production (Song et al., 2015; Pang and Zhang, 2019; Gupta et al., 2021). The aim of this chapter is to provide an overview of AM's sustainable approach to zero-waste manufacturing and cleaner production. It will address the basic concepts, issues, and opportunities related to the use of AM for sustainable manufacturing, drawing on previous research to highlight key advances and achievements in this

DOI: 10.1201/9781003309123-1

field. The first part of this chapter introduces the concept of zero-waste manufacturing, which prevents the generation of waste at source and encourages the use of already utilized resources and energy to create closed-loop systems. The principles of zero-waste production, including waste prevention, reduction, and diversion, are explained with their potential benefits for long-term environmental, economic, and social sustainability (Zotos et al., 2009). Maximizing resource utilization (raw material utilization), reducing pollution, and increasing the productivity of the manufacturing process are the main goals of cleaner production. Compared to the conventional manufacturing process, AM permits the precise deposition of material exactly where it is needed, resulting in less material waste (Sanchez-Rexach et al., 2020; Camacho et al., 2018; Mehrpouya et al., 2019). With AM, complex geometries can be designed, and multiple parts can be combined into a single component with less material and weight, requiring less energy during manufacturing and transportation (Huang et al., 2013). AM opens the application for manufacturing of recycled or bio-based materials, which can further increase the sustainability of the manufacturing process (Reichert et al., 2020).

The challenges of combining AM with sustainable manufacturing have been discussed in detail in this chapter. These challenges include the lack of standardized techniques for recycling and disposal of AM materials, their high cost and limited supply, the energy-intensive nature of some AM processes, and the potential risks to one's health and safety when handling AM materials. This chapter also discusses the increasing demands and challenges related to AM in different areas, including research and development. Due to zero-waste materials, AM is the most demanding manufacturing process to protect our environment and reduce the quantity of pollutants as compared to traditional/conventional manufacturing processes.

This study also explains the future applications of AM in sustainable manufacturing. This study suggests the creation of new sustainable materials that are especially suited for AM, advancing circular economy principles in AM processes, integrating AM with other advanced manufacturing technologies such as robotics and artificial intelligence to maximize resource utilization, and establishing standardized recycling and disposal procedures for AM materials. To enhance the use of AM for sustainable manufacturing on a larger scale, this chapter will also emphasize the significance of cooperative efforts among academics, industry, and policymakers. In attaining zero-waste manufacturing and greener production, AM employs a sustainable strategy.

1.2 PRINCIPLES OF ZERO-WASTE MANUFACTURING AND CLEANER PRODUCTION

1.2.1 Basic Principles of Zero-Waste Manufacturing and Cleaner Production

Cleaner production is a concept that encompasses many of the operational practicalities of best environmental practices on a mine site. It aims at maximizing resource usage and operational efficiency during the production of minerals. The concept also extends to minimize waste disposal and rehabilitation requirements, and its application is linked to continuous improvement in environmental and economic performance. It is an integrated and preventive approach to minimize environmental risk rather than a curative approach. The benefits of cleaner production can include less waste, recovery of valuable by-products, improved environmental performance, increased productivity, better efficiency, reduced energy consumption, and an overall reduction in costs. The main aim of this work of cleaner production and zero-waste manufacturing is to minimize waste production, lessen the impact on the environment, and increase the sustainability of industrial operations (Song et al., 2015; Glavic and Lukman, 2007). These principles serve as a framework for the creation and application of policies and procedures that emphasize resource preservation, waste reduction, and pollution abatement. The fundamental principle of zero-waste manufacturing utilizing a circular economy is shown in Figure 1.1. Here are the key principles of zero-waste manufacturing and cleaner production.

The environmental effect of a product or process is assessed holistically by zero-waste manufacturing and cleaner production throughout a product's complete life cycle. This principle calls for considering how a product or procedure will affect the environment from the time that raw materials are extracted through manufacture, usage, and disposal or recycling. It is feasible to find possibilities for waste reduction and environmental improvement at every step of the product's life cycle by adopting a life cycle approach (Bhander et al., 2003). The secret to reaching zero waste is waste prevention. The goal is to prevent the generation of waste in the first place by developing processes and products that minimize waste generation—for example, through product design, process optimization, and waste reduction methods (Walton et al., 1998). This concept is about reducing pollution and the environmental impact of production processes. It involves the introduction of technologies, practices, and policies that limit or eliminate the discharge of hazardous chemicals, pollutants, and waste into the environment (Zhang et al., 1997). Pollution control devices, process optimization, and the use of environmentally friendly materials are examples of pollution prevention strategies. Efficient management and utilization of resources such as raw materials, electricity, and water are key to waste-free production and more environmentally friendly production (Hu et al., 2011).

To minimize waste and its impact on the environment, the AM process generates less waste material compared to other conventional manufacturing processes. The main component of zero-waste manufacturing and clean production involves stakeholders such as workers, suppliers, consumers, and local communities. Developing principles for zero-waste and clean production helps to increase support, awareness, and commitment (Clay et al., 2007; Veleva et al., 2017). It supports the concept of closing the material flow, where waste is seen as a valuable resource that can be recycled or used instead of being thrown away. Circular economy practices such as recycling, reuse, and recovery can help reduce waste and the need for new raw materials (Romero-Hernández and Romero, 2018). Zero-waste manufacturing and more environmentally friendly production is a continuous development. Constant monitoring, measurement, and evaluation are important to track success and find areas for improvement.

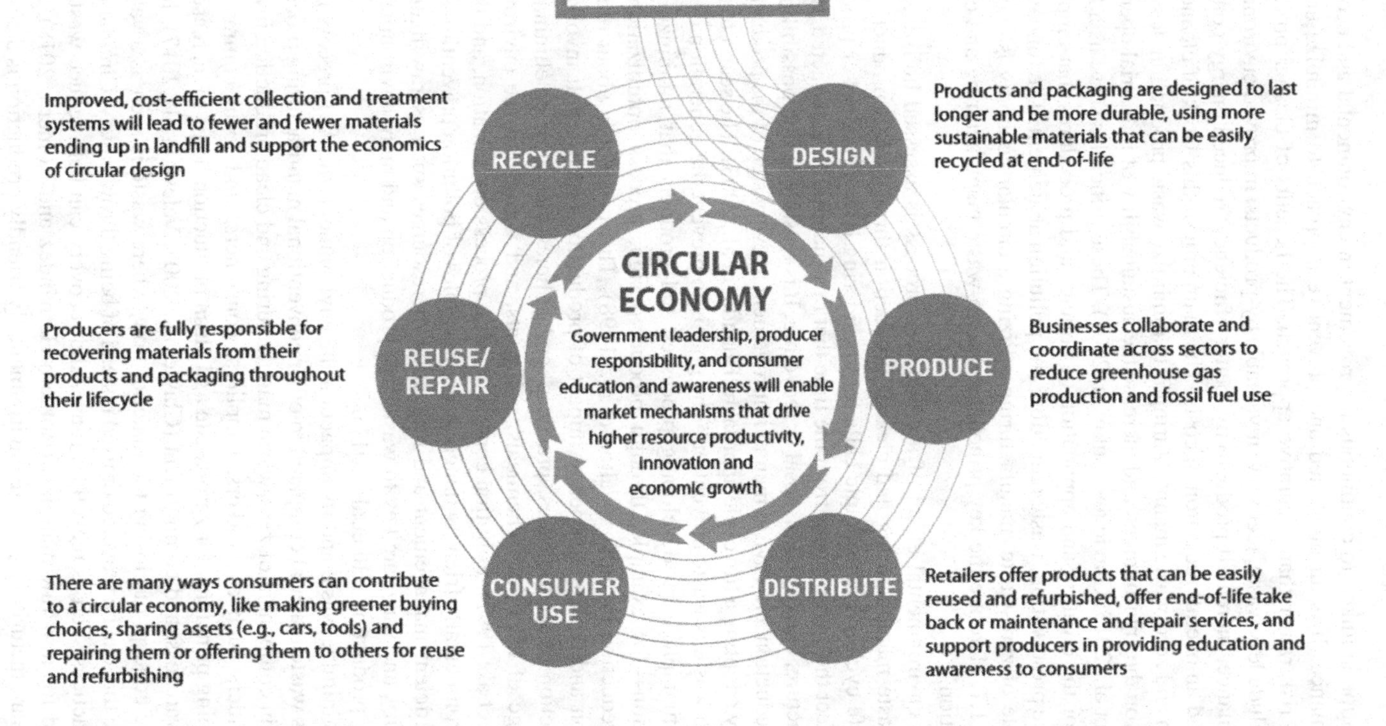

FIGURE 1.1 Zero-waste manufacturing using circular economy.

Source: Ontario's Strategy for a Waste-Free Ontario: Building the Circular Economy (2023). https://www.ontario.ca/page/strategy-waste-free-ontario-building-circular-economy. Used under CC BY-4.0.

1.2.2 Zero-Waste Manufacturing and Cleaner Production Based on AM

The manufacturing sector may see a transformation because of the disruptive technology known as AM. As with any manufacturing process, it's critical to think about how AM affects the environment and work toward waste-free manufacturing and cleaner production (Stavropoulos et al., 2020; Gopal et al., 2022). Figure 1.2 depicts the systematic flow circular flow cycle of zero-waste production based on AM. Following are some guidelines that might help AM develop sustainable practices.

AM trash output may be greatly decreased by putting recycling and reuse practices into place. For instance, discarded or unsuccessful prints may be recycled and repurposed as a source of material for subsequent printing. Reusing and recycling products can reduce trash sent to landfills and help preserve natural resources (Javaid et al., 2021). Designing items designed for AM is one of the tenets of zero-waste manufacturing. Design for additive manufacturing (DfAM) is the process of enhancing a product's design to capitalize on the special characteristics of AM, such as its capacity to produce complicated geometries and combine many components into a single component. Less material is wasted during the printing process when components are designed to be optimized for AM, which reduces material consumption and waste production (Yang and Zhao, 2015; Gibson et al., 2021).

A complete life cycle analysis (LCA) of the AM process may assist in pinpointing opportunities for improvement and direct decision-making toward more environmentally friendly procedures. To determine the potential for waste reduction and cleaner production, LCA entails assessing the environmental effects of AM from raw material extraction through end-of-life disposal (Peng et al., 2018). Carefully choosing and using materials in AM is another crucial idea. AM may lessen its negative effects on the environment by using sustainable and green resources like recycled feedstock and biodegradable polymers. Zero-waste

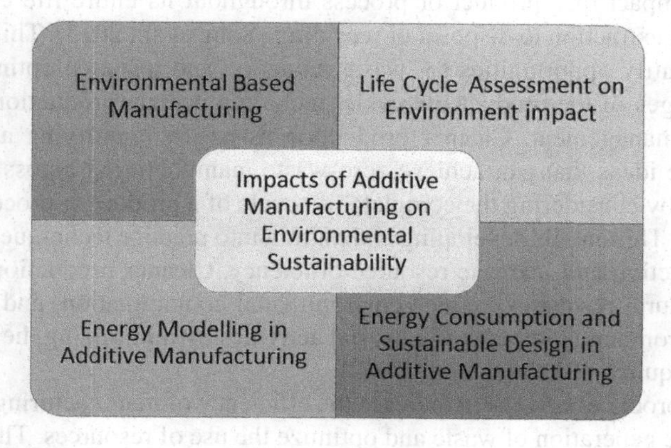

FIGURE 1.2 Flow chart of zero-waste manufacturing based on AM.

manufacturing may also be achieved by maximizing material consumption and reducing overhangs, supports, and extra material in prints (Rett et al., 2021; Overcash, 1996). In AM, post-processing and finishing activities like polishing or painting can lead to waste production and resource consumption. Zero-waste manufacturing may be achieved by using sustainable post-processing and finishing practices, such as employing eco-friendly finishing materials and reducing trash output. The environmental effect of AM is significantly influenced by energy use. Consequently, optimizing energy use throughout the printing process can help produce goods more sustainably. The use of energy-efficient printers, print parameter optimization, and the use of renewable energy sources to power the printing process are just a few ways to do this (Bourhis et al., 2013). It is critical for the successful implementation of AM to educate and train the personnel on sustainable practices. Educating operators on environmentally beneficial practices like correct material utilization, recycling, and energy optimization may aid in the promotion of zero-waste manufacturing and cleaner production in AM.

Finally, integrating sustainable practices in AM is critical for zero-waste manufacturing and cleaner production. The AM industry can make significant strides toward a more sustainable and environmentally responsible manufacturing process by considering DfAM principles, material selection and usage, recycling and reusing, energy efficiency, post-processing, and finishing, conducting LCA, and providing education and training.

1.3 ROLE OF CLEANER PRODUCTION IN ACHIEVING ZERO-WASTE MANUFACTURING

Cleaner production is a crucial approach that complements the goal of achieving zero-waste manufacturing. It involves optimizing resource utilization, reducing pollution, and improving the efficiency of manufacturing processes to minimize the environmental impact of production while maximizing economic benefits (Stavropoulos et al., 2020). It takes a life cycle perspective, considering the environmental impact of a product or process throughout its entire life cycle, from raw material extraction to disposal or recycling (Song et al., 2015). This approach helps to identify opportunities for waste reduction and resource optimization at different stages of the product life cycle, including design, production, use, and end-of-life management. Cleaner production assists in identifying and putting into practice ideas that can achieve zero-waste manufacturing across the whole value chain by considering the complete life cycle of a product or process (Zhang et al., 1997). This entails developing and putting into practice techniques to reduce waste production and increase resource efficiency. Cleaner production serves to preserve natural resources, reduce environmental contamination, and lessen the overall environmental impact of industrial activities by minimizing the number of resources required (Bhander et al., 2003).

Cleaner production seeks to increase the efficiency of manufacturing processes to reduce the generation of waste and optimize the use of resources. This includes introducing new technologies, redesigning processes, and optimizing processes to reduce resource waste, material losses, and energy consumption. By improving process efficiency, cleaner production reduces waste, uses fewer resources, and increases overall productivity and profitability (Almeida et al., 2013). Reducing resource consumption, waste production, and pollution control costs can lead to financial savings. By maximizing resource utilization and improving process efficiency, clean production can increase the competitiveness and sustainability of manufacturing operations (Hegab et al., 2023). This has a positive impact on the economy, as it increases profitability, reduces the risk of noncompliance, and strengthens brand reputation. By identifying and tackling the primary sources of pollution in industrial processes, cleaner production focuses on preventing pollution at its source (Camacho et al., 2018). By removing pollution at its source, decreasing the need for expensive waste treatment and disposal, and safeguarding the environment and human health, cleaner production lessens the environmental impact of industrial operations.

For achieving zero-waste manufacturing, cleaner production plays a critical role in increasing resource utilization, reducing pollution, and boosting process efficiency. It encourages environmentally friendly production methods that also preserve resources and boost economic gains. Manufacturing industries must adopt AM to achieve zero-waste manufacturing by implementing cleaner production techniques, which will help to promote a more sustainable and environmentally friendly method of production.

1.4 ROLE OF CLEANER PRODUCTION FOR ZERO-WASTE MANUFACTURING BY AM

AM provides various advantages over conventional manufacturing processes, including decreased material waste, better design flexibility, and enhanced sustainability due to its capacity to construct complex and customized things layer by layer.

A concept known as zero-waste manufacturing strives to completely eradicate waste production throughout the whole manufacturing process, from the extraction of raw materials to the final disposal. Conventional manufacturing processes frequently produce large amounts of waste, such as scrap materials, surplus inventories, and rejected goods, which contribute to resource depletion and environmental damage (Hegab et al., 2023). AM holds the promise as a means of reaching zero-waste production since it can reduce or even eliminate several of these waste streams. Through its capacity to maximize material consumption, AM may play a significant role in zero-waste production. Materials must frequently be chopped, machined, and shaped in conventional manufacturing processes, which results in considerable material waste (Camacho et al., 2018). AM, on the other hand, constructs items layer by layer, adding material only where it is required, reducing material waste. When compared to typical production processes, this can result in material savings of up to 90%. AM allows for the use of recycled or bio-based materials, lowering the need for virgin resources and reducing waste related to raw material extraction.

Due to it being easy to use and easy to manufacture very difficult parts as compared to conventional manufacturing techniques, AM gives more design freedom. Manufacturing products with optimized geometry that reduces the need for extra material (waste materials) can result in the more effective use of materials. For instance, internal lattice systems can be used in lightweight constructions to conserve material while retaining structural integrity (Sanchez-Rexach et al., 2020). This design flexibility also makes it possible to combine numerous parts into a single component, lowering the overall component count, streamlining the supply chain, and eliminating the waste produced during the assembly process. The flexibility of AM to provide on-demand production is another way it may aid in zero-waste manufacturing. Mass production is a common need in traditional manufacturing, which leads to excess inventory and waste from overproduction (Mehrpouya et al., 2019). By using AM, products are manufactured very easily and quickly in small batches and on-demand, which eliminates the need for additional inventory and cuts down on waste from overproduction (Camacho et al., 2018). Since goods may be manufactured close to where they will be used, less wastage is associated with transportation.

AM can help zero-waste manufacturing in addition to decreasing waste during the production phase. Conventional manufacturing costs are a little higher for single-piece product fabrication as compared to the AM process, which easily fulfills the product customization to match unique consumer demands, minimizing waste and the manufacture of unnecessary goods. AM makes it possible to repair broken or worn-out items by simply printing replacement components, hence increasing product lifespans and decreasing waste from product obsolescence (Zotos et al., 2009). AM may significantly contribute to zero-waste production by allowing the circular economy idea. To minimize waste and maximize resource utilization, the circular economy is an economic model that encourages the reuse, renovation, remanufacturing, and recycling of goods and resources. By making it possible to produce spare parts locally, lowering the need to replace complete items, and prolonging their lifespan, AM promotes the ideas of the circular economy (Despeisse et al., 2017). It makes it easier to produce goods by using recycled or reused materials, lowering the need for virgin resources, and fostering resource conservation. The capability of AM to cut energy use is a crucial benefit in attaining zero-waste production. The significant machining, cutting, and shaping of materials used in traditional manufacturing processes can be energy intensive. As just the material required for the object being printed is used, AM often utilizes less energy.

1.5 ACHIEVEMENTS AND ADVANCEMENTS OF AM FOR SUSTAINABLE MANUFACTURING

According to a literature study, it has been observed that over the years, examining the successes and developments in the field of AM for sustainable manufacturing gives better output in different areas and also in research and innovation. The creation of bio-based and recycled materials for AM processes is a noteworthy achievement in the field of AM for sustainable manufacturing (Pakkanen et al., 2017). With the development of bioplastics that have been produced from renewable resources like algae, soybean oil, or corn flour, researchers have made great progress toward developing sustainable materials that may be utilized in 3D printers. These materials provide a renewable and biodegradable substitute to conventional plastics derived from petroleum, as well as reduce the environmental effect of AM procedures (Zotos et al., 2009). Sustainable production through AM has also benefited from improvements in design optimization and material use (Gupta et al., 2021). To optimize the design for minimal material consumption while retaining structural integrity, researchers have created sophisticated computational tools that allow for complicated geometries and lattice structures (Du Plessis et al., 2019). As a result, there is now less material waste, less energy used, and more effective AM processes, which leads to manufacturing that is both sustainable and affordable.

Manufacturing has been made possible by AM, which also lessens the need for transportation and the accompanying carbon emissions. Local fabrication of replacement parts, tools, and prototypes may be carried out on-site as 3D printers become more widely available and reasonably priced, removing the need for long-distance transportation and lowering the overall carbon footprint of manufacturing (Sasson and Johnson, 2016). The utilization of recycled and upcycled materials is an important development in AM for sustainable production. According to a review of the literature, it has been observed that the fabrication of products using recycled metals, polymers, and other materials as feedstock for AM techniques reduces waste and also decreases the dependency on virgin raw materials (Gupta et al., 2021). By using waste materials for useful goods, this strategy not only supports the ideals of the circular economy but also lessens the environmental effect of production. AM has achieved major strides in environmentally friendly product design. Lots of researchers and academicians have created inventive designs that maximize product usefulness, durability, and resource efficiency by using the flexibility and modification possibilities of AM (Haghnegahdar et al., 2022). These features, which are not feasible with conventional production techniques, include lightweight constructions, optimized part consolidation, and complicated geometries, leading to products with lower material consumption and higher sustainability performance.

AM has also been used in the creation of green energy solutions. AM has been used to design complex geometries for renewable energy components, including solar panels, wind turbine blades, and energy storage devices. These enhanced designs have increased longevity, decreased energy consumption, and better performance, encouraging the use of sustainable energy technology (Fraţila and Rotaru, 2017). Substantial developments and advancements have been made in the field of AM for sustainable manufacturing. The creation of bio-based and recycled materials, improvements in design optimization and material utilization, localized manufacturing, the use of recycled and upcycled materials, sustainable product design, and implementations of sustainable energy solutions are a few examples of these accomplishments (Hegab et al., 2023). The acceptance of AM as a sustainable manufacturing technique is being sparked by these successes, which also encourage resource efficiency, cut waste, and lessen the environmental effect of production operations. Continued study in this discipline shows promise for new developments in the industry and can help manufacture a more sustainable future.

1.6 SUMMARY

Despite the potential benefits of AM for zero-waste manufacturing and cleaner production, several problems and limits must be solved. One of the difficulties is the high energy consumption of various AM techniques, particularly those that employ high-temperature procedures such as selective laser sintering or electron beam melting. Because the energy needs for these processes might counterbalance the material waste reduction obtained by AM, careful consideration of energy sources and efficiency is required to ensure sustainable results. Based on this chapter, it has been obtained that AM is one of the fast-growing manufacturing technologies that has zero raw materials wastage as compared to conventional manufacturing technology. AM also has the potential to make a substantial contribution to these objectives. Zero-waste manufacturing and cleaner production are essential strategies for attaining sustainability in manufacturing. AM may help promote sustainable manufacturing practices by minimizing material waste, optimizing material utilization, lowering energy consumption, and facilitating the use of eco-friendly materials. To make sure that AM is employed in sustainable manufacturing, various issues and constraints must be resolved.

REFERENCES

Almeida CM, Bonilla SH, Giannetti BF, Huisingh D. Cleaner Production initiatives and challenges for a sustainable world: An introduction to this special volume. *Journal of Cleaner Production*. 2013;47:1.

Bhander GS, Hauschild M, McAloone T. Implementing life cycle assessment in product development. *Environmental Progress*. 2003;22(4):255–67.

Bourhis FL, Kerbrat O, Hascoet JY, Mognol P. Sustainable manufacturing: Evaluation and modeling of environmental impacts in additive manufacturing. *The International Journal of Advanced Manufacturing Technology*. 2013;69:1927–39.

Camacho DD, Clayton P, O'Brien WJ, Seepersad C, Juenger M, Ferron R, Salamone S. Applications of additive manufacturing in the construction industry–A forward-looking review. *Automation in Construction*. 2018;89:110–9.

Clay S, Gibson D, Ward J. Sustainability Victoria: Influencing resource use, towards zero waste and sustainable production and consumption. *Journal of Cleaner Production*. 2007;15(8–9):782–6.

Despeisse M, Baumers M, Brown P, Charnley F, Ford SJ, Garmulewicz A, Knowles S, Minshall TH, Mortara L, Reed-Tsochas FP, Rowley J. Unlocking value for a circular economy through 3D printing: A research agenda. *Technological Forecasting and Social Change*. 2017;115:75–84.

Du Plessis A, Broeckhoven C, Yadroitsava I, Yadroitsev I, Hands CH, Kunju R, Bhate D. Beautiful and functional: A review of biomimetic design in additive manufacturing. *Additive Manufacturing*. 2019;27:408–27.

Fraţila D, Rotaru H. Additive manufacturing–a sustainable manufacturing route. In *MATEC web of conferences 2017* (2017, Vol. 94, p. 03004). EDP Sciences.

Gibson I, Rosen D, Stucker B, Khorasani M, Gibson I, Rosen D, Stucker B, Khorasani M. Design for additive manufacturing. *Additive Manufacturing Technologies*. 2021:555–607.

Glavic P, Lukman R. Review of sustainability terms and their definitions. *Journal of Cleaner Production*. 2007;15(18):1875–85.

Gopal M, Lemu HG, Gutema EM. Sustainable additive manufacturing and environmental implications: Literature review. *Sustainability*. 2022;15(1):504.

Gupta H, Kumar A, Wasan P. Industry 4.0, cleaner production and circular economy: An integrative framework for evaluating ethical and sustainable business performance of manufacturing organizations. *Journal of Cleaner Production*. 2021;295:126253.

Gupta K, Laubscher RF, Davim JP, Jain NK. Recent developments in sustainable manufacturing of gears: A review. *Journal of Cleaner Production*. 2016;112:3320–30.

Haghnegahdar L, Joshi SS, Dahotre NB. From IoT-based cloud manufacturing approach to intelligent additive manufacturing: Industrial Internet of Things—An overview. *The International Journal of Advanced Manufacturing Technology*. 2022:1–8.

Hegab H, Khanna N, Monib N, Salem A. Design for sustainable additive manufacturing: A review. *Sustainable Materials and Technologies*. 2023; 35:e00576.

Hu J, Xiao Z, Zhou R, Deng W, Wang M, Ma S. Ecological utilization of leather tannery waste with circular economy model. *Journal of Cleaner Production*. 2011;19(2–3):221–8.

Huang SH, Liu P, Mokasdar A, Hou L. Additive manufacturing and its societal impact: A literature review. *The International Journal of Advanced Manufacturing Technology*. 2013;67:1191–203.

Javaid M, Haleem A, Singh RP, Suman R, Rab S. Role of additive manufacturing applications towards environmental sustainability. *Advanced Industrial and Engineering Polymer Research*. 2021;4(4):312–22.

Kurniawan TA, Avtar R, Singh D, Xue W, Othman MH, Hwang GH, Iswanto I, Albadarin AB, Kern AO. Reforming MSWM in Sukunan (Yogjakarta, Indonesia): A case-study of applying a zero-waste approach based on circular economy paradigm. *Journal of Cleaner Production*. 2021;284:124775.

Mehrpouya M, Dehghanghadikolaei A, Fotovvati B, Vosooghnia A, Emamian SS, Gisario A. The potential of additive manufacturing in the smart factory industrial 4.0: A review. *Applied Sciences*. 2019;9(18):3865.

Ngo TD, Kashani A, Imbalzano G, Nguyen KT, Hui D. Additive manufacturing (3D printing): A review of materials, methods, applications and challenges. *Composites Part B: Engineering*. 2018;143:172–96.

Ontario's Strategy for a Waste-Free Ontario: Building the Circular Economy. Available online: https://www.ontario.ca/page/strategy-waste-free-ontario-building-circular-economy (accessed on 17 September 2023).

Overcash M. Cleaner production: basic principles and development. *Clean Technology*. 1996;2(1):1–6.

Pakkanen J, Manfredi D, Minetola P, Iuliano L. About the use of recycled or biodegradable filaments for sustainability of 3D printing: State of the art and research opportunities. Sustainable Design and Manufacturing 2017: Selected papers on Sustainable Design and Manufacturing 4. 2017:776–85.

Pang R, Zhang X. Achieving environmental sustainability in manufacture: A 28-year bibliometric cartography of green manufacturing research. *Journal of Cleaner Production*. 2019;233:84–99.

Peng T, Kellens K, Tang R, Chen C, Chen G. Sustainability of additive manufacturing: An overview on its energy demand and environmental impact. *Additive Manufacturing*. 2018;21:694–704.

Pietzsch N, Ribeiro JL, de Medeiros JF. Benefits, challenges and critical factors of success for Zero Waste: A systematic literature review. *Waste Management*. 2017;67:324–53.

Reichert CL, Bugnicourt E, Coltelli MB, Cinelli P, Lazzeri A, Canesi I, Braca F, Martínez BM, Alonso R, Agostinis L, Verstichel S. Bio-based packaging: Materials, modifications, industrial applications and sustainability. *Polymers*. 2020;12(7):1558.

Rett JP, Traore YL, Ho EA. Sustainable materials for fused deposition modeling 3D printing applications. *Advanced Engineering Materials*. 2021;23(7):2001472.

Romero-Hernández O, Romero S. Maximizing the value of waste: From waste management to the circular economy. *Thunderbird International Business Review*. 2018;60(5):757–64.

Sanchez FA, Boudaoud H, Camargo M, Pearce JM. Plastic recycling in additive manufacturing: A systematic literature review and opportunities for the circular economy. *Journal of Cleaner Production*. 2020;264:121602.

Sanchez-Rexach E, Johnston TG, Jehanno C, Sardon H, Nelson A. Sustainable materials and chemical processes for additive manufacturing. *Chemistry of Materials*. 2020;32(17):7105–19.

Sasson A, Johnson JC. The 3D printing order: variability, supercenters and supply chain reconfigurations. *International Journal of Physical Distribution & Logistics Management*. 2016;46:82–94.

Singh S, Ramakrishna S, Gupta MK. Towards zero waste manufacturing: A multidisciplinary review. *Journal of Cleaner Production*. 2017;168:1230–43.

Song Q, Li J, Zeng X. Minimizing the increasing solid waste through zero waste strategy. *Journal of Cleaner Production*. 2015;104:199–210.

Stavropoulos P, Papacharalampopoulos A, Tzimanis K, Lianos A. Manufacturing resilience during the coronavirus pandemic: On the investigation manufacturing processes agility. *European Journal of Social Impact and Circular Economy*. 2020;1(3):28–57.

Veleva V, Bodkin G, Todorova S. The need for better measurement and employee engagement to advance a circular economy: Lessons from Biogen's "zero waste" journey. *Journal of Cleaner Production*. 2017;154:517–29.

Walton SV, Handfield RB, Melnyk SA. The green supply chain: Integrating suppliers into environmental management processes. *International Journal of Purchasing and Materials Management*. 1998;34(1):2–11.

Yang S, Zhao YF. Additive manufacturing-enabled design theory and methodology: A critical review. *The International Journal of Advanced Manufacturing Technology*. 2015;80:327–42.

Zhang HC, Kuo TC, Lu H, Huang SH. Environmentally conscious design and manufacturing: A state-of-the-art survey. *Journal of Manufacturing Systems*. 1997;16(5):352–71.

Zotos G, Karagiannidis A, Zampetoglou S, Malamakis A, Antonopoulos IS, Kontogianni S, Tchobanoglous G. Developing a holistic strategy for integrated waste management within municipal planning: Challenges, policies, solutions and perspectives for Hellenic municipalities in the zero-waste, low-cost direction. *Waste Management*. 2009;29(5):1686–92.

2 Sustainable 3D Printing-Based Smart Solution for the Maintenance of Heritage Structures

Vinay Kumar
University Center of Research and Development,
Chandigarh University, Mohali, India

2.1 INTRODUCTION

In the past two decades, some studies have reported the harmful effects of global environmental changes on heritage structures that pointed toward the fast degradation of valuable cultural heritage sites/articles and the need for the conservation of heritage structures (Viles and Cutler, 2012; De la Torre, 2013). Some researchers have outlined the values-based sustainable approach for the management and conservation of heritage structures that may include the analysis of geological degradation in the heritage sites due to fragility and natural vulnerability (Poulios, 2010; García-Ortiz et al., 2014). The novel color reference charts reported in recent studies may also be used to monitor and validate the degradation of cultural heritage (Ramírez Barat et al., 2021). The conventional approach of heritage conservation practices has been observed as a useful approach for the preservation of human heritage that includes structures, cathedrals, tombs, masonry, etc. (Montgomery et al., 2020). It has been reported that sustainable solutions for heritage conservation may play a vital role in stabilizing the economic conditions of developing countries (Salameh et al., 2022). Advanced digital technologies like digitized oral history, complex reality, virtual reality, and 3D modeling have also been reported as innovative heritage conservation tools for sustainable and long-term preservation of world heritage (Matusiak et al., 2017; Fernández-Palacios et al., 2017; Rodriguez Echavarria et al., 2012). With an increase in demand for sustainable construction solutions, many researchers have explored strategic frameworks to develop smart and sustainable solutions for the rehabilitation of heritage structures (Du Plessis, 2007). Some studies have indicated that smart or

DOI: 10.1201/9781003309123-2

advanced functional materials with programmable properties possess the capability to conserve cultural heritage articles like ancient stones, clothes, paintings, etc. (Baglioni et al., 2021). The investigations performed on the re-evaluation of heritage waste to prepare sustainable repair solutions for the same heritage (by reusing the debris of the particular heritage site and reconstructing the repair material) outlined that this approach may be used for better harmonization between deteriorating heritage and weathering effect (Ross, 2020; Keshtkaran, 2011). Also, efficient town planning has been reported as a significant factor in the creation of effective and sustainable industrial heritage for future generations (Guo et al., 2021). The literature survey (as per the Scopus database) on sustainable construction materials for heritage conservation has outlined 7,419 documents reported in the past 30 years. Figure 2.1 shows the graphical representation indicating the gradual rise in the number of documents reported since 1989 for this particular research area of sustainability.

Sustainable development solutions play a vital role in balancing environmental degradation. Therefore the need for such solutions is highly required in present times to preserve the grand heritage for coming generations (Cassar, 2009). The interconnectivity of tourism planning, urban heritage places, reconciling conservation practices, and sustainable development approaches may be regarded as an effective approach to obtaining sustainable solutions for heritage structures (Nasser, 2003). Some studies have outlined the advantages of merging environmental conservation and historic preservation to draw suitable practice routes for building sustainable management solutions for heritage sites and cultural heritage (Leifeste and Stiefel, 2018; Sodangi et al., 2014; Kayan, 2019). Figure 2.2 shows

FIGURE 2.1 Research publications in the area of sustainable construction materials for the conservation of cultural heritage and heritage structures.

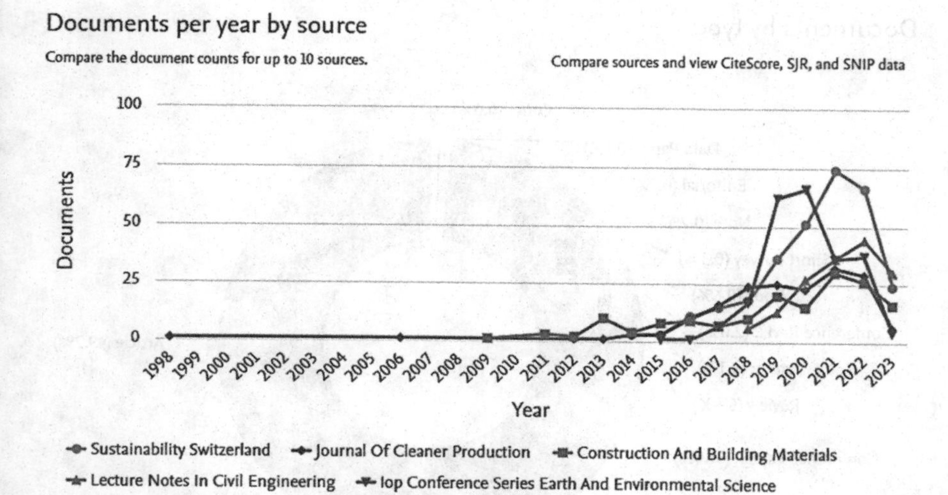

FIGURE 2.2 Number of documents published by various resources in the past 30 years on sustainable construction solutions.

the documents published by various sources in the past 30 years on sustainable solution development, highlighting that significant changes have been observed in reported documents for the investigations of smart and sustainable construction solutions for civil engineering-based cleaner and eco-friendly building and construction materials.

As evident from Figure 2.2, the number of documents reported on sustainability increased exponentially in the past five years. This shows that the application of advanced construction and manufacturing practices largely affects the energy demands due to the Industrial Revolution. The introduction of additive manufacturing in the industrial and construction sector boosted the economy through the fabrication of sustainable consumer products at the industrial level and sustainable buildings like low-cost houses and schools in developing nations at the infrastructure building level (Singh and Kumar, 2022; Peng et al., 2018). The manufacturing practices like fused filament fabrication have been highlighted as a novel method to utilize waste plastic materials to develop new construction materials like composites for sustainable construction (Javaid et al., 2021). Based on such road maps of sustainable additive manufacturing, Figure 2.3 shows the various categories of documents like research articles and chapters published on the 3D printing of polymer composite matrix for structural engineering applications (Kumar and Czekanski, 2018). This indicates that the 3D printing techniques possess a large potential in the implementation of sustainable construction solutions also for heritage conservation by reframing the construction methods toward ecological balance and circular economy (Anastasiadou and Vettese, 2019).

The categorization of various published documents outlines that laser additive manufacturing and fused filament fabrication, in comparison to conventional

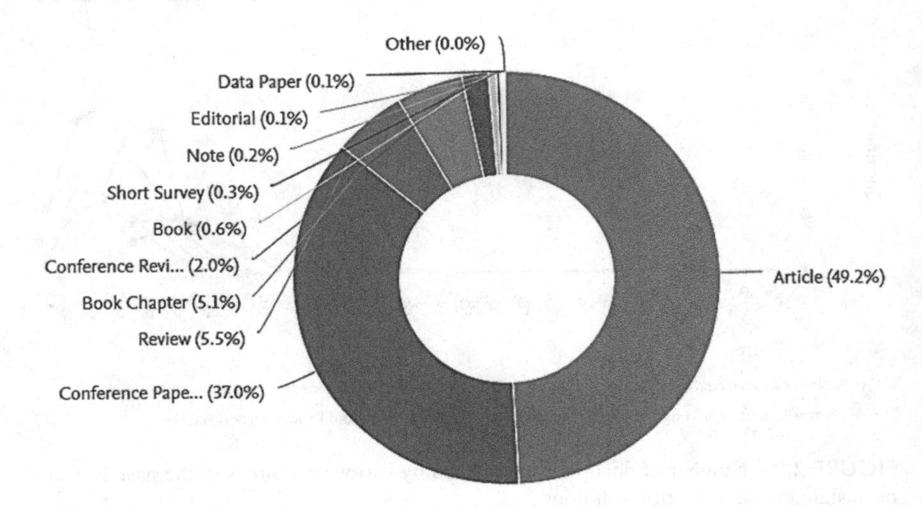

FIGURE 2.3 Distribution of various categories of research documents published on sustainable construction solutions (as per Scopus database).

machining practices, have contributed in a much better way to sustainability assessment (Jiang et al., 2019; Kumar et al., 2022a). Besides engineering, sustainable construction solutions have been reported by researchers having environmental studies backgrounds and material science and social science backgrounds (Jiang and Fu, 2023; Kumar et al., 2022b). Figure 2.4 shows the distribution of reported literature on sustainable materials from various subject areas. Some recent studies reported on 3D printing have highlighted the usefulness of sustainable manufacturing of 3D printable multi-materials to control carbon emissions in industrial sectors. Therefore, such manufacturing practices may be incorporated for sustainable industrialization (Agrawal and Vinodh, 2021; Singh et al., 2019, 2020; Kumar et al., 2022c).

The literature reported on the fabrication of composite structures for industrial and structural applications has outlined the advantages of recycling waste plastic like polyamide (PA6), polyvinylidene fluoride (PVDF), polylactic acid (PLA), acrylonitrile butadiene styrene (ABS), low-density polyethylene (LDPE), etc., for the preparation of sustainable consumer products. Such products have wide acceptability for 4D printing applications in biomedical engineering also (Singh et al., 2018; Singh, 2022). Alongside sustainable materials for construction/structural applications, some studies have reported the development of customizable repair and maintenance solutions for heritage structures. The thermoplastic composite matrix based such solutions that have been derived from the debris of the same heritage site possess sensing capabilities also that may be utilized for heritage building information modeling applications (Kumar et al., 2023a, 2023b).

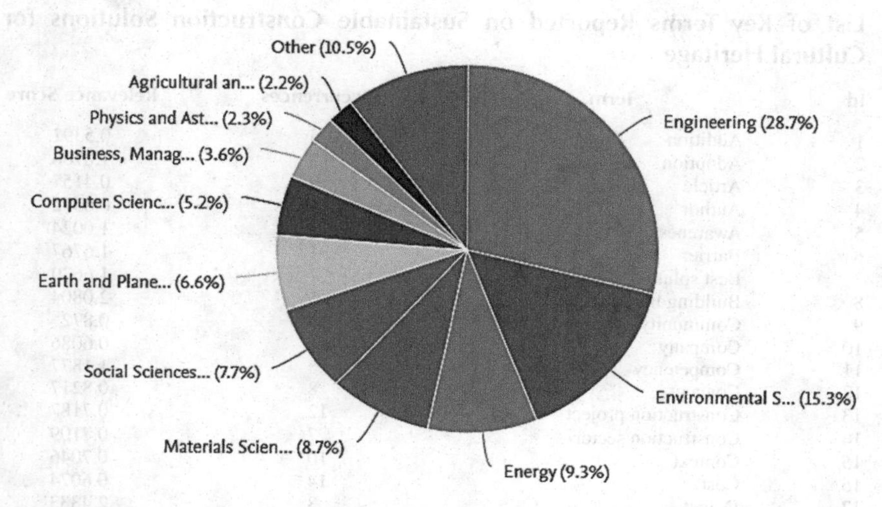

FIGURE 2.4 Subject area-wise distribution of documents reported on sustainable construction materials.

Nevertheless, very little work is reported on investigations of sustainability in the developed construction materials, repair, and maintenance solutions for longer preservation and conservation of heritage structures.

2.2 RESEARCH GAP AND PROBLEM FORMULATION

Based on the literature review, the Scopus database was analyzed to obtain the research gap in the area of sustainable solution fabrication for heritage structures. The database was processed on the VOS viewer open-source software package to club the various research areas in which sustainable solutions have been proposed for construction applications. In 7,419 articles, a total of 77 highly reported key terms were outlined by keeping the number of the most relevant key terms = 5. Out of these 77 terms, 60% most relevant terms are listed in Table 2.1, which outlines the highly reported studies on sustainable solutions for cultural heritage and heritage structures.

Based on Table 2.1, Figure 2.5 shows the web of keywords interlinking the various research areas that outline the studies reported on sustainability, sustainable construction, and sustainable solutions for engineering applications. Since the heritage structures reflect the ancient architectural talent of humans in the present world, this motivates conservators to preserve such structures and sites for future generations.

The present scenario demands the preservation/conservation of heritage structures and cultural heritage by utilizing sustainable solutions. Utilization of recycled plastic waste for the preparation of hybrid materials with smart and

TABLE 2.1
List of Key Terms Reported on Sustainable Construction Solutions for Cultural Heritage

Id	Term	Occurrences	Relevance Score
1	Addition	5	0.5391
2	Adoption	12	1.0768
3	Article	9	0.4157
4	Author	6	0.6372
5	Awareness	7	1.0024
6	Barrier	31	1.6767
7	Best solution	7	1.6639
8	Building material	5	2.0804
9	Community	10	0.872
10	Company	11	0.6086
11	Competency	5	1.1877
12	Concept	8	0.8217
13	Construction project	12	0.7187
14	Construction sector	7	0.7109
15	Context	10	0.7046
16	Cost	14	0.6074
17	Defect	8	2.4833
18	Energy	9	0.4653
19	Engineer	6	0.6708
20	Energy storage	6	1.5883
21	Field	7	0.8933
22	Implementation	7	0.696
23	Importance	9	0.4771
24	Innovation	8	1.0423
25	Interview	10	0.3783
26	Lack	12	1.2206
27	Major barrier	5	1.6014
28	Model	14	0.5198
29	Order	11	0.3549
30	Organization	8	0.6185
31	Problem	11	0.4523
32	Requirement	9	0.3274
33	Role	15	0.675
34	SDGs	9	1.4058
35	Singapore	8	1.1736
36	Small contractor	8	2.0772
37	Survey	13	0.2294
38	Sustainability	27	0.5387
39	Sustainable building	11	1.9964
40	Sustainable construction	55	0.4551
41	Sustainable development goal	8	1.5042
42	Sustainable solution	9	0.7373
43	TBL	5	1.1472
44	Term	8	0.7486
45	Topic	7	0.6315
46	Training	5	0.7507
47	Type	6	2.7014
48	University	5	1.4956
49	Use	21	0.5885
50	Wood waste	6	2.0303

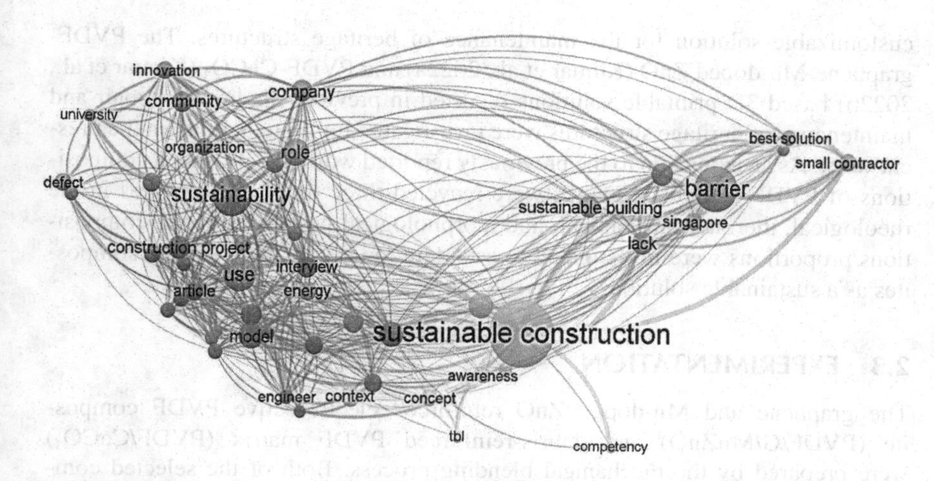

FIGURE 2.5 Web of keywords outlining interconnectivity of various research areas.

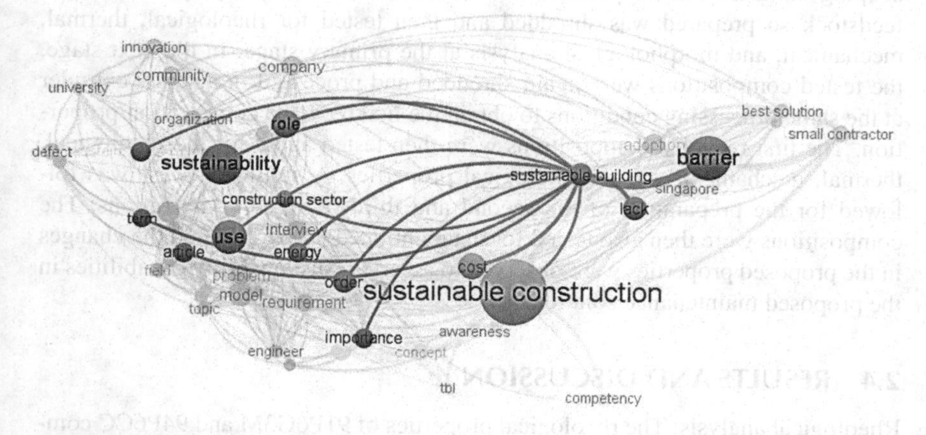

FIGURE 2.6 Research gap highlighting the missing link between maintenance of defects in heritage buildings by sustainable solutions.

customizable properties may be incorporated to obtain sustainable solutions for the maintenance of heritage structures (Kumar et al., 2022d, 2022e, 2022f). But hitherto, very little has been reported on sustainable 3D printing–based smart solutions for the maintenance of heritage structures. Figure 2.6 highlighted the gap in the literature that barriers to achieving sustainable heritage buildings lie in the suitable addressing of defects and damages faced by heritage structures due to weathering, biodegradation, and earthquakes. This study highlights the sustainability properties of the PVDF thermoplastic composite matrix-based customizable solution for the maintenance of heritage structures. The PVDF-graphene-Mn-doped ZnO (Kumar et al., 2022a) and PVDF-$CaCO_3$ (Kumar et al., 2022b) based 3D printable solutions reported in previous studies for repair and maintenance of heritage structures were investigated for sustainability in the present work. As an extension to the previously reported work, the proposed compositions of PVDF composite matrix were recycled three times and changes in the rheological, thermal, mechanical, and morphological properties of the compositions/proportions were investigated to ascertain the acceptability of the composites as a sustainable solution.

2.3 EXPERIMENTATION

The graphene and Mn-doped ZnO reinforced electro-active PVDF composite (PVDF/G/MnZnO) and debris-reinforced PVDF matrix (PVDF/$CaCO_3$) were prepared by the mechanical blending process. Both of the selected compositions/proportions were 91%PVDF/6%G/3%MnZnO (91P6G3M) and 94%PVDF/6%$CaCO_3$ (94P6CC). The compositions were processed on a single screw extruder (Make: Felfil Evo, Italy) to prepare the feedstock filament (by keeping the extrusion temperature 220°C and extrusion rate 6rpm). The sample feedstock so prepared was shredded and then tested for rheological, thermal, mechanical, and morphological analysis at the primary stage. In the next stage, the tested compositions were again shredded and processed in a screw extruder at the same processing conditions to obtain the first recycled composition/proportion. The first recycled compositions were then tested again for the rheological, thermal, mechanical, and morphological properties. A similar process was followed for the preparation of the second and third recycled compositions. The compositions were then also tested for the mentioned properties, and the changes in the proposed properties were observed to ascertain sustainability capabilities in the proposed maintenance solution.

2.4 RESULTS AND DISCUSSION

Rheological analysis: The rheological properties of 91P6G3M and 94P6CC composites (comprising of melt flow index (MFI), viscosity, and density properties) at the initial stage and corresponding stages of first, second, and third recycling. The proposed MFI testing was performed as per ASTM D1238 by using PACORR MFI tester (Make: Ghaziabad, India). Figure 2.7 shows the MFI setup used, and Table 2.2 shows the results obtained for the rheological properties of 91P6G3M and 94P6CC.

The results obtained for MFI, density, and viscosity of 91P6G3M and 94P6CC composites highlighted that the rheological properties of the PVDF thermoplastic composite matrix reduced slightly after every recycling stage. This may be because every heating stage for melting the polymer matrix composite contributed

FIGURE 2.7 MFI test setup used for rheological testing.

TABLE 2.2
Observations for Rheological Properties of Recycled PVDF Composites

Composition/ Proportion	Recycling Stage	MFI (g/10 min)	Density (g/cm³)	Viscosity (Pa-s)
91P6G3M	Initial (I)	2.7 ± 0.1	1.5 ± 0.2	4409.3 ± 0.5
	1st(1)	2.65 ± 0.3	1.45 ± 0.5	4405.4 ± 0.2
	2nd(2)	2.63 ± 0.2	1.42 ± 0.3	4389.2 ± 0.4
	3rd(3)	2.59 ± 0.3	1.41 ± 0.4	4385.7 ± 0.1
94P6CC	Initial (I)	2.44 ± 0.2	1.641 ± 0.5	12500.5 ± 0.3
	1st(1)	2.41 ± 0.1	1.635 ± 0.1	12441.8 ± 0.4
	2nd(2)	2.38 ± 0.4	1.629 ± 0.3	12435.2 ± 0.5
	3rd(3)	2.37 ± 0.2	1.628 ± 0.4	12431.7 ± 0.3

Note: Composites prepared at each stage were tested 03times, and the average value was recorded.

to decreasing the molecular strength of the monomer chain. Concerning 3D printing applications, no significant degradation in the rheological properties of composites was observed, as the average MFI obtained was 2.6 g/10 min for 91P6G3M and 2.4 g/10 min for 94P6CC. Therefore, the composites may be used for acceptable 3D printing applications. Figure 2.8 shows the trends obtained for the rheological properties of 91P6G3M and 94P6CC composites.

Thermal analysis: The thermal properties of 91P6G3M and 94P6CC composites, like heat capacity, glass transition temperature, and crystallinity, were investigated by differential scanning calorimetry (DSC) (Make: Mettler Toledo DSC 3 StarE system, Switzerland) at the initial stage and corresponding stages of first,

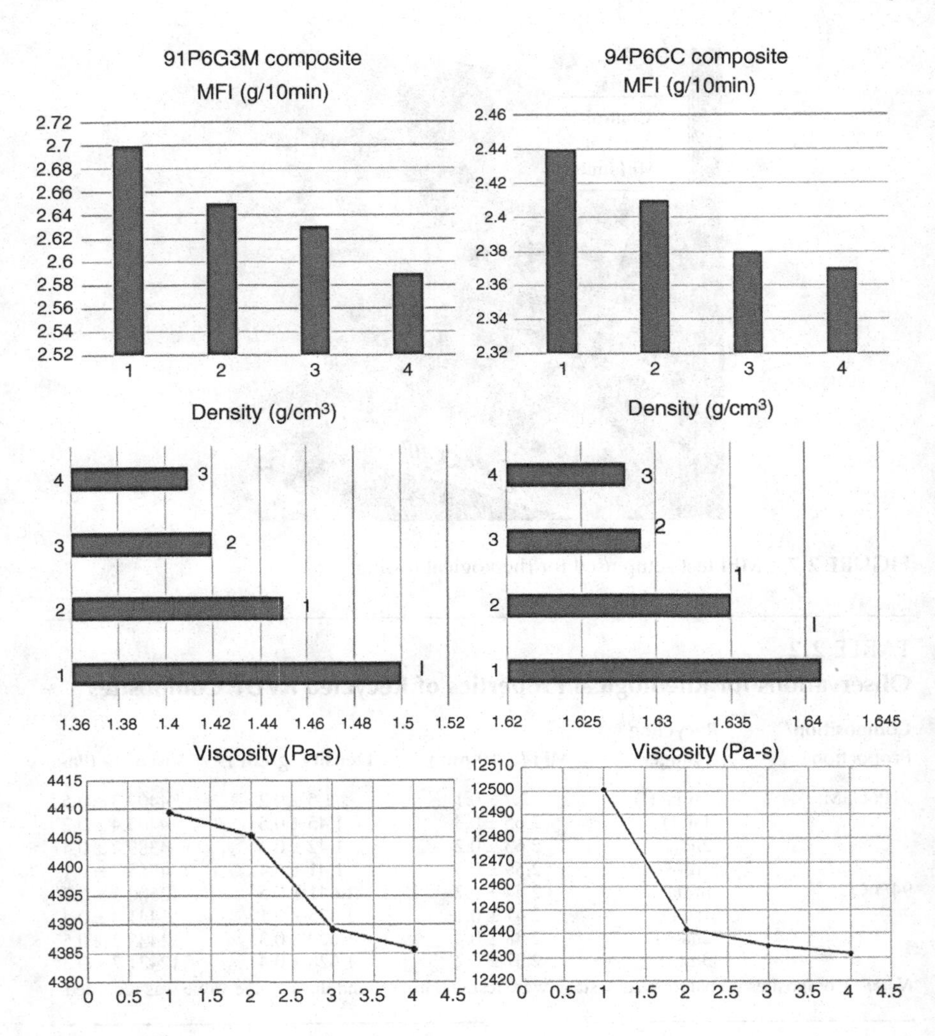

FIGURE 2.8 Trends obtained for rheological properties of PVDF composites.

second, and third recycling. Figure 2.9 shows the DSC setup used, and Table 2.3 shows the results obtained for the thermal properties of 91P6G3M and 94P6CC. The changes observed in the thermal properties of the composites with successive heating and cooling cycles (during DSC testing), and extrusion heating/melting (during filament fabrication) are represented in the form of graphs shown in Figure 2.10.

The results obtained for heat capacity, crystallinity, and glass transition temperature of 91P6G3M and 94P6CC composites highlighted that the thermal properties of the PVDF thermoplastic composite matrix decreased in close range for successive recycling stages (i.e., from the initial to the third recycling stage). This

FIGURE 2.9 DSC test setup used for thermal testing.

TABLE 2.3
Observations for Thermal Properties of Recycled PVDF Composites

Composition/ Proportion	Recycling Stage	Heat Capacity (J/g)	Crystallinity (%)	Glass Transition Temperature (T_g) (°C)
91P6G3M	Initial	85.8 ± 0.4	57 ± 0.3	−35.15 ± 0.2
	1st	84.6 ± 0.5	56 ± 0.2	−35.11 ± 0.1
	2nd	82.7 ± 0.2	55 ± 0.4	−35.09 ± 0.5
	3rd	81.8 ± 0.3	54.5 ± 0.1	−35.12 ± 0.3
94P6CC	Initial	58.9 ± 0.4	59 ± 0.4	−35.10 ± 0.6
	1st	57.3 ± 0.2	57 ± 0.5	−35.18 ± 0.2
	2nd	56.1 ± 0.5	55.8 ± 0.3	−35.16 ± 0.4
	3rd	55.8 ± 0.1	54.3 ± 0.6	−35.11 ± 0.5

may be because every heating and cooling stage for DSC analysis contributed to the refinement of the polymer matrix chain. With every recycling stage, the impurities present in the compositions decreased significantly and finally decreased the molecular strength of the monomer chain. In regard to 3D printing applications, no significant degradation in the thermal properties of composites was observed, as the average heat capacity obtained for 91P6G3M was 83.5 J/g and 56.5 J/g for 94P6CC. Therefore, the composites may be used for acceptable 3D printing applications. Figure 2.10 shows the trends obtained for thermal properties (in terms of heat capacity) of 91P6G3M and 94P6CC composites.

Mechanical properties: The mechanical properties like peak strength (S_{peak}), break strength (S_{break}), modulus of toughness (G), and Young's modulus (E) of 91P6G3M and 94P6CC composites were investigated by universal testing machine (UTM) (Make: Shanta Engineers, Pune, India) for extruded filament

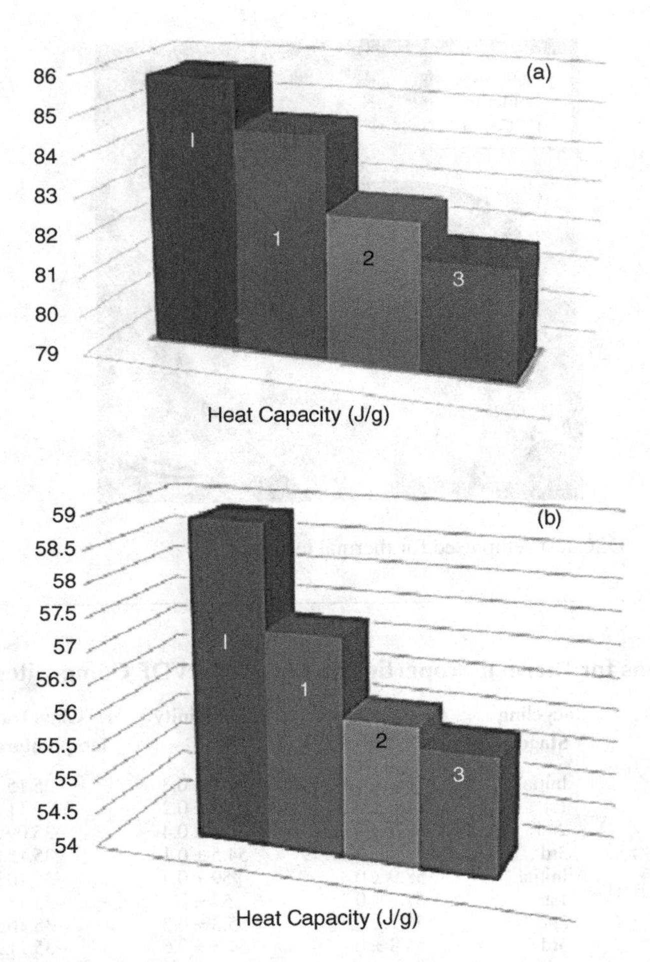

FIGURE 2.10 Trends obtained for thermal properties of 91P6G3M (a) and 94P6CC composites (b).

samples of the initial stage and corresponding stages of the first, second, and third recycling. Figure 2.11 shows the UTM setup used, and Table 2.4 shows the results obtained for mechanical properties of 91P6G3M and 94P6CC. The changes observed in the mechanical properties of the composites are presented in the form of graphs (as shown in Figure 2.12).

The results obtained for the mechanical strength of 91P6G3M and 94P6CC composites highlighted that the S_{peak}, S_{break}, E, and G properties of the PVDF thermoplastic composite matrix deceased for successive recycling stages—i.e., from the initial to the third recycling stage. This may be because the decrease in the bond strength of the material led to the decrease in the ultimate tensile strength of the composites. With every single recycling stage, the bond strength and reduced

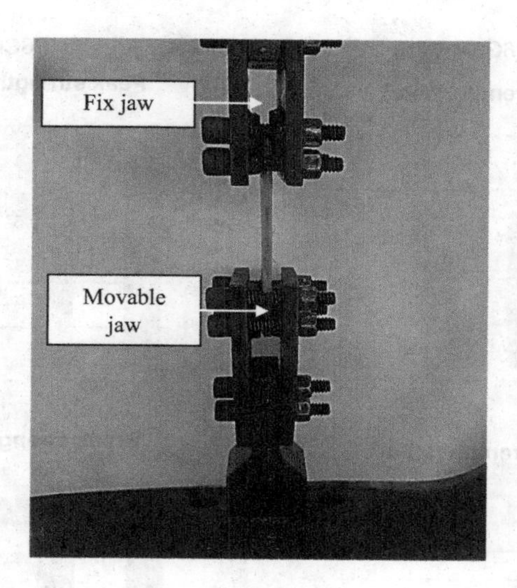

FIGURE 2.11 DSC test setup used for thermal testing.

TABLE 2.4
Observations for Mechanical Properties of Recycled PVDF Composites

Composition/ Proportion	Recycling Stage	S_{peak} (MPa)	S_{break} (MPa)	G (MPa)	E (MPa)
91P6G3M	Initial	25.9 ± 0.4	22.5 ± 0.2	0.85 ± 0.1	380 ± 0.4
	1st	24.8 ± 0.6	22.02 ± 0.3	0.84 ± 0.3	378 ± 0.5
	2nd	24.1 ± 0.2	21.7 ± 0.2	0.83 ± 0.5	377 ± 0.3
	3rd	23.6 ± 0.4	21.6 ± 0.1	0.81 ± 0.2	374 ± 0.1
94P6CC	Initial	26.1 ± 0.3	24.4 ± 0.4	0.87 ± 0.3	375 ± 0.2
	1st	25.8 ± 0.5	24.2 ± 0.5	0.86 ± 0.1	373 ± 0.4
	2nd	25.4 ± 0.3	23.7 ± 0.1	0.84 ± 0.4	371 ± 0.3
	3rd	24.9 ± 0.1	23.6 ± 0.3	0.83 ± 0.2	369 ± 0.5

slighter than in the earlier stage, and hence, the E and G of the composite decreased. As regards 3D printing applications, again, no significant degradation in the mechanical properties of composites was observed, as the average E obtained for 91P6G3M was 377 MPa and 373 MPa for 94P6CC. Therefore, the composites may be used for acceptable 3D printing applications. Figure 2.12 shows the trends obtained for mechanical properties (in terms of peak and break strength) of 91P6G3M and 94P6CC composites.

Morphological analysis: the morphological properties of the 91P6G3M and 94P6CC composite samples were also investigated at all four stages. The photomicrographs of filament samples were obtained (using Toolmaker's microscope) at ×30 magnification along the cross-section (to ascertain the porosity (P %)) and longitudinal axis (to ascertain the surface roughness (Ra)). Table 2.5 shows the

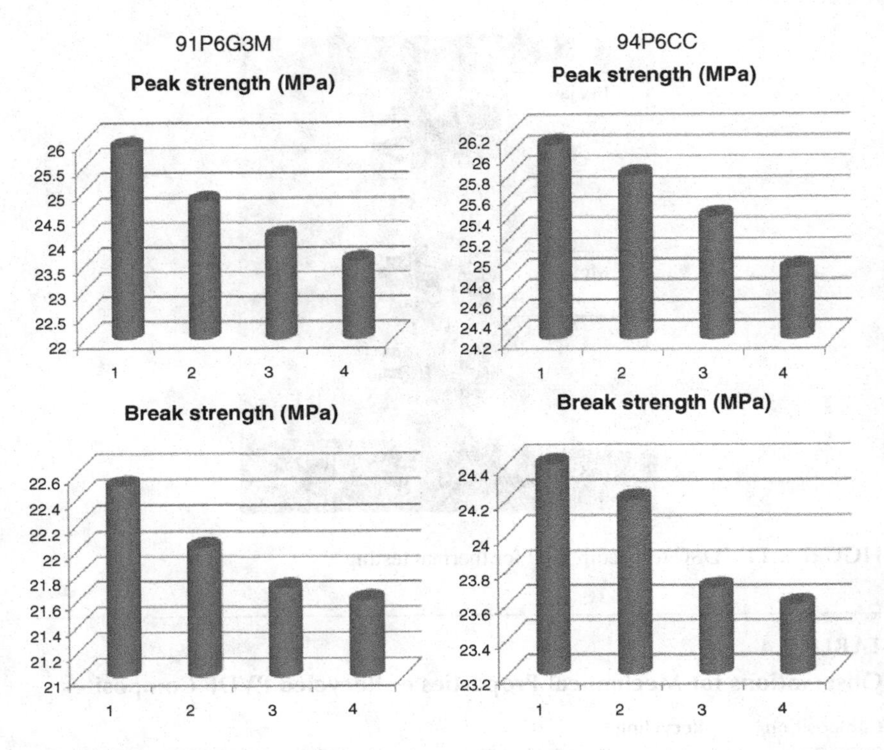

FIGURE 2.12 Trends obtained for peak strength and break strength of 91P6G3M and 94P6CC composites.

TABLE 2.5
Observations for Morphological Properties of PVDF Composites

Composition/ Proportion	Recycling Stage	Porosity (%)	Ra (nm)
91P6G3M	Initial	6.35	55.21
	1st	7.17	58.33
	2nd	8.46	61.67
	3rd	9.87	74.02
94P6CC	Initial	7.28	57.19
	1st	8.54	59.27
	2nd	11.49	62.76
	3rd	12.65	66.58

results obtained for the morphological properties of the composites. Figure 2.13 shows the photomicrographs obtained for porosity and roughness analysis of the compositions/proportions.

The morphological analysis outlined that 91P6G3M and 94P6CC composites possess acceptable P% (9.78% in 91P6G3M, 12.65% in 94P6CC along

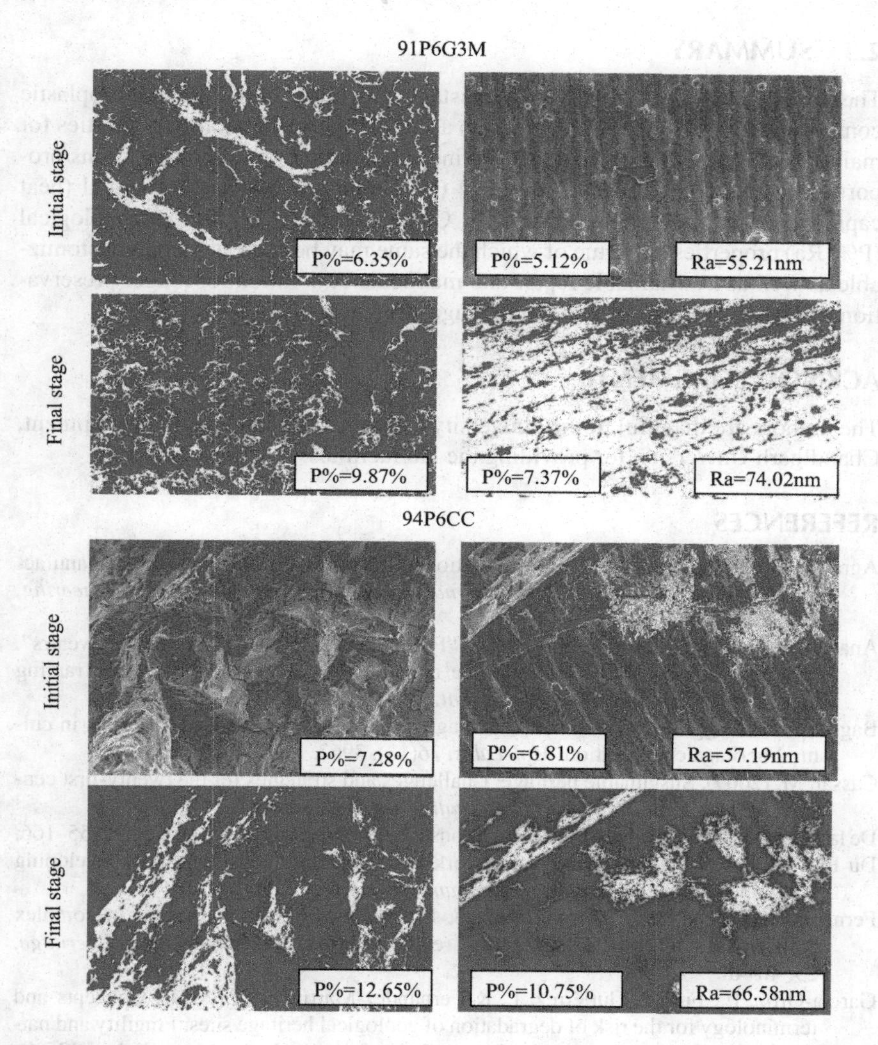

FIGURE 2.13 Photomicrographs observations for 91P6G3M and 94P6CC composite filament samples.

cross-section) and Ra (74.02 nm in 91P6G3M, 66.58 nm in 94P6CC along the transverse axis) for sustainable 3D printing applications, as controlled surface characteristics were obtained for the proposed polymer matrix composite after three stages of recycling. The stability in the rheological, thermal, mechanical, and morphological properties of 91P6G3M and 94P6CC composites indicated that the proposed compositions are suitable sustainable solutions for 3D printing applications.

2.5 SUMMARY

The investigation performed on multistage recycling of PVDF thermoplastic composites (91P6G3M and 94P6CC) to ascertain the sustainability properties for maintenance of heritage structures outlined that the proposed compositions/proportions possess controlled rheological (MFI, density, viscosity), thermal (heat capacity, crystallinity), mechanical (E, G, S_{peak}, and S_{break}) and morphological (P%, Ra) properties by virtue of which the same may be used as a smart customizable as well as a sustainable repair and maintenance solution for longer preservation of nonstructural cracks in the heritage structures.

ACKNOWLEDGMENT

The authors are thankful to the University Centre of Research and Development, Chandigarh University, for providing the lab facilities.

REFERENCES

Agrawal, R., & Vinodh, S. (2021). Prioritisation of drivers of sustainable additive manufacturing using best worst method. *International Journal of Sustainable Engineering*, *14*(6), 1587–1603.

Anastasiadou, C., & Vettese, S. (2019). "From souvenirs to 3D printed souvenirs". Exploring the capabilities of additive manufacturing technologies in (re)-framing tourist souvenirs. *Tourism Management*, *71*, 428–442.

Baglioni, M., Poggi, G., Chelazzi, D., & Baglioni, P. (2021). Advanced materials in cultural heritage conservation. *Molecules*, *26*(13), 3967.

Cassar, M. (2009). Sustainable heritage: Challenges and strategies for the twenty-first century, APT bulletin. *Journal of Preservation Technology*, *40*(1), 3–11.

De la Torre, M. (2013). Values and heritage conservation. *Heritage & Society*, *6*(2), 155–166.

Du Plessis, C. (2007). A strategic framework for sustainable construction in developing countries. *Construction Management and Economics*, *25*(1), 67–76.

Fernández-Palacios, B. J., Morabito, D., & Remondino, F. (2017). Access to complex reality-based 3D models using virtual reality solutions. *Journal of cultural heritage*, *23*, 40–48.

García-Ortiz, E., Fuertes-Gutiérrez, I., & Fernández-Martínez, E. (2014). Concepts and terminology for the risk of degradation of geological heritage sites: Fragility and natural vulnerability, a case study. *Proceedings of the Geologists' Association*, *125*(4), 463–479.

Guo, P., Li, Q., Guo, H., & Li, H. (2021). Quantifying the core driving force for the sustainable redevelopment of industrial heritage: Implications for urban renewal. *Environmental Science and Pollution Research*, *28*(35), 48097–48111.

Javaid, M., Haleem, A., Singh, R. P., Suman, R., & Rab, S. (2021). Role of additive manufacturing applications towards environmental sustainability. *Advanced Industrial and Engineering Polymer Research*, *4*(4), 312–322.

Jiang, J., & Fu, Y. F. (2023). A short survey of sustainable material extrusion additive manufacturing. *Australian Journal of Mechanical Engineering*, *21*(1), 123–132.

Jiang, Q., Liu, Z., Li, T., Cong, W., & Zhang, H. C. (2019). Emergy-based life-cycle assessment (Em-LCA) for sustainability assessment: A case study of laser additive manufacturing versus CNC machining. *The International Journal of Advanced Manufacturing Technology*, *102*, 4109–4120.

Kayan, B. A. (2019). Sustainable built heritage: Maintenance management appraisal approach. *Journal of Cultural Heritage Management and Sustainable Development*, *9*, 266–281

Keshtkaran, P. (2011). Harmonization between climate and architecture in vernacular heritage: A case study in Yazd, Iran. *Procedia Engineering*, *21*, 428–438.

Kumar, S., & Czekanski, A. (2018). Roadmap to sustainable plastic additive manufacturing. *Materials Today Communications*, *15*, 109–113.

Kumar, V., Singh, R., & Ahuja, I. S. (2022a). On 3D printing of electro-active PVDF-Graphene and Mn-doped ZnO nanoparticle-based composite as a self-healing repair solution for heritage structures. *Proceedings of the Institution of Mechanical Engineers, Part B: Journal of Engineering Manufacture*, *236*(8), 1141–1154.

Kumar, V., Singh, R., & Ahuja, I. S. (2022b). Secondary recycled polyvinylidene–limestone composite in 4D printing applications for heritage structures: Rheological, thermal, mechanical, spectroscopic, and morphological analysis. *Proceedings of the Institution of Mechanical Engineers, Part E: Journal of Process Mechanical Engineering*, 09544089221104771.

Kumar, V., Singh, R., & Ahuja, I. S. (2022c). On Correlation of Rheological, Thermal, Mechanical and Morphological Properties of Mechanically Blended PVDF-Graphene Composite for 4d Applications.

Kumar, V., Singh, R., & Ahuja, I. S. (2022d). Tertiary recycling of plastic solid waste for additive manufacturing. In Rupinder Singh, Ranvijay Kumar (Eds.), *Additive Manufacturing for Plastic Recycling* (pp. 93–109). CRC Press.

Kumar, V., Singh, R., & Ahuja, I. S. (2022e). Hybrid feedstock filament processing for the preparation of composite structures in heritage repair. In Rupinder Singh, Ranvijay Kumar (Eds.), *Additive Manufacturing for Plastic Recycling* (pp. 159–170). CRC Press.

Kumar, V., Singh, R., & Ahuja, I. S. (2022f). In-house Development of Smart Materials for 4D Printing. In Rupinder Singh (Ed.), *4D Imaging to 4D Printing* (pp. 85–102). CRC Press.

Kumar, V., Singh, R., & Ahuja, I. S. (2023a). Multi-material printing of PVDF composites: A customized solution for maintenance of heritage structures. *Proceedings of the Institution of Mechanical Engineers, Part L: Journal of Materials: Design and Applications*, *237*(3), 554–564.

Kumar, V., Singh, R., & Ahuja, I. S. (2023b). On debris reinforced-PVDF, composite-based 3D printed sensors for restoration of heritage building. *National Academy Science Letters*, *46*, 329–332.

Leifeste, A., & Stiefel, B. L. (2018). *Sustainable heritage: Merging environmental conservation and historic preservation*. Routledge.

Matusiak, K. K., Tyler, A., Newton, C., & Polepeddi, P. (2017). Finding access and digital preservation solutions for a digitized oral history project: A case study. *Digital Library Perspectives*, *33*(2), 88–99.

Montgomery, R. A., Borona, K., Kasozi, H., Mudumba, T., & Ogada, M. (2020). Positioning human heritage at the center of conservation practice. *Conservation Biology*, *34*(5), 1122–1130.

Nasser, N. (2003). Planning for urban heritage places: Reconciling conservation, tourism, and sustainable development. *Journal of Planning Literature*, *17*(4), 467–479.

Peng, T., Kellens, K., Tang, R., Chen, C., & Chen, G. (2018). Sustainability of additive manufacturing: An overview on its energy demand and environmental impact. *Additive Manufacturing*, *21*, 694–704.

Poulios, I. (2010). Moving beyond a values-based approach to heritage conservation. *Conservation and management of Archaeological Sites*, *12*(2), 170–185.

Ramírez Barat, B., Cano, E., Molina, M. T., Barbero-Álvarez, M. A., Rodrigo, J. A., & Menéndez, J. M. (2021). Design and validation of tailored color reference charts for monitoring cultural heritage degradation. *Heritage Science*, *9*, 1–9.

Rodriguez Echavarria, K., Kaminski, J., & Arnold, D. (2012). 3D heritage on mobile devices: scenarios and opportunities. In *Progress in Cultural Heritage Preservation: 4th International Conference, EuroMed 2012, Limassol, Cyprus, October 29–November 3, 2012. Proceedings 4* (pp. 149–158). Berlin Heidelberg: Springer.

Ross, S. M. (2020). Re-evaluating heritage waste: Sustaining material values through deconstruction and reuse. *The Historic Environment: Policy & Practice*, *11*(2–3), 382–408.

Salameh, M. M., Touqan, B. A., Awad, J., & Salameh, M. M. (2022). Heritage conservation as a bridge to sustainability assessing thermal performance and the preservation of identity through heritage conservation in the Mediterranean city of Nablus. *Ain Shams Engineering Journal*, *13*(2), 101553.

Singh, R. (Ed.). (2022). *4D Imaging to 4D Printing: Biomedical Applications* (1st ed.). CRC Press. https://doi.org/10.1201/9781003205531

Singh, R., & Kumar, R. (Eds.). (2022). *Additive Manufacturing for Plastic Recycling: Efforts in Boosting A Circular Economy* (1st ed.). CRC Press. https://doi.org/10.1201/9781003184164

Singh, R., Dureja, J. S., Dogra, M., Gupta, M. K., Jamil, M., & Mia, M. (2020). Evaluating the sustainability pillars of energy and environment considering carbon emissions under machining ofTi-3Al-2.5 V. *Sustainable Energy Technologies and Assessments*, *42*, 100806.

Singh, R., Kumar, R., Farina, I., Colangelo, F., Feo, L., & Fraternali, F. (2019). Multi-material additive manufacturing of sustainable innovative materials and structures. *Polymers*, *11*(1), 62.

Singh, R., Kumar, R., Ranjan, N., Penna, R., & Fraternali, F. (2018). On the recyclability of polyamide for sustainable composite structures in civil engineering. *Composite Structures*, *184*, 704–713.

Sodangi, M., Khamdi, M. F., Idrus, A., Hammad, D. B., & Ahmed Umar, A. (2014). Best practice criteria for sustainable maintenance management of heritage buildings in Malaysia. *Procedia Engineering*, *77*, 11–19.

Viles, H. A., & Cutler, N. A. (2012). Global environmental change and the biology of heritage structures. *Global Change Biology*, *18*(8), 2406–2418.

3 Sustainability in Manufacturing Industries

Emergence in Industry 4.0

Ketan Badogu, Khushwant Kour, and Ranvijay Kumar
Chandigarh University, Mohali, India

3.1 INTRODUCTION

Industry 4.0 (4IR) is a new trend centered on automated intelligence technologies. In this new era, the use of current manufacturing abilities within the factors of incorporating breakthrough an innovative information technology is very vital for economic viability in the marketplace. As shown in Figure 3.1, 4IR allows cyberspace and physical structures to profitably collaborate in order to develop intelligent manufacturing facilities by rethinking the role of people (Dilberoglu et al., 2017). Its primary concepts connected with the virtual world include the Internet of Things (IoT), big data (BD), cloud computing, and further developments, while its physical domain encompasses automated robotic manufacturing and additive manufacturing (AM) (Liu & Xu, 2017).

The term "4IR" is nowadays an industrial buzzword. In recent years, organizations have sought to incorporate sustainability considerations into their business operations (Luthra et al., 2017; Mangla et al., 2015; Sarkis & Zhu, 2018). Moreover, industries are battling to meet customers' ever-changing demands while still maintaining their environmental sustainability (Stock & Seliger, 2016). Industrial leaders are implementing current innovations such as 3D printing (3DP), IoT, data analytics, and 4IR to foster a creative company environment. Cyber-physical systems encompass the concept of IoT, which involves acquiring data from physical objects through computer networks or fast wireless connections. The data collected from various sources like products, machinery, and manufacturing processes generates a substantial amount of statistical information that needs to be transmitted and analyzed. Additionally, data from design documents, customer orders, supplier deliveries, inventory, and logistics also contribute to this extensive dataset. This large and diverse collection of information is commonly known as BD, which holds significant significance within the context of the 4IR (Wang & Wang, 2016).

DOI: 10.1201/9781003309123-3

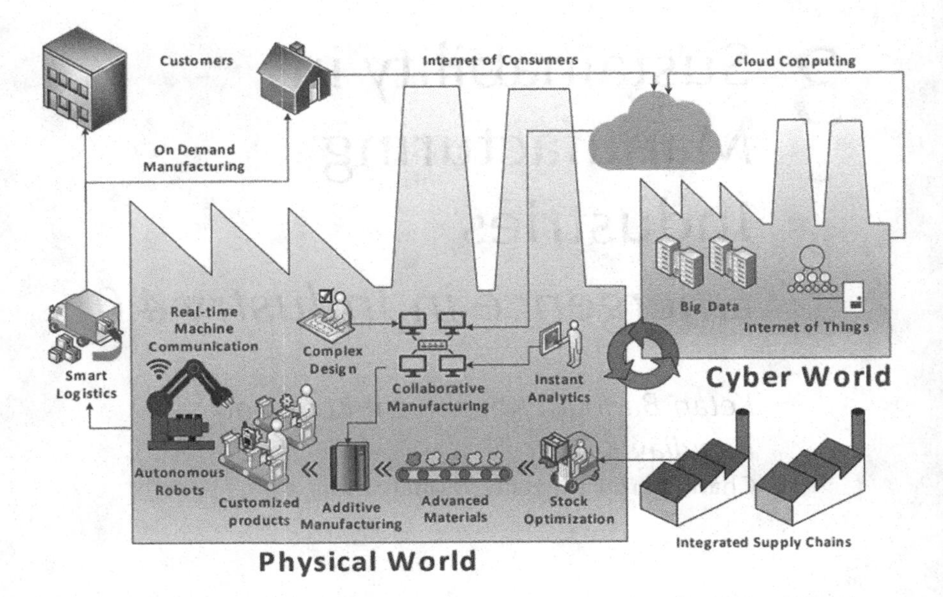

FIGURE 3.1 Conceptual mapping of smart factories system with general characteristics necessary for 4IR.

Source: Dilberoglu, U. M., Gharehpapagh, B., Yaman, U., & Dolen, M. (2017). The role of additive manufacturing in the era of industry 4.0. *Procedia Manufacturing*, *11*(June), 545–554. Reproduced with permission.

The capabilities of traditional production technologies restrict the physical component of automated manufacturing facilities. This limitation emphasizes the significance of AM within the context of the 4IR. To meet the demand for mass customization in the 4IR era, it is essential to develop nontraditional production methods. With its ability to fabricate complex items featuring advanced attributes, like new materials and unique shapes, AM holds immense potential as a prominent technology for manufacturing customized products. With improvements in product quality, AM is increasingly being utilized across diverse industries such as aerospace, biomedical, and manufacturing (Thompson et al., 2016). The aforementioned technologies, particularly 4IR, are drastically changing supply chain management (SCM) behavior. The 4IR-based sustainable development-oriented concept assists business leaders in not only incorporating environmental safeguards and management initiatives but also in coupling safety concerns, such as resource utilization, staff members as well as community well-being, and more advanced and versatile procedures, into their supply chains (Tjahjono et al., 2017).

The 4IR accelerates industrial development, but it also threatens the long-term viability of present industrial systems (Hermann et al., 2016). This may lead to more ecological imbalances on Earth, such as increased utilization of resources, global warming, issues related to a change in climate, and greater energy demand. Furthermore, expanding industrialization adds to the deterioration of worker health and safety. In this context, industrial production systems that use current technology must be ecologically, financially, and socially balanced (Wang et al., 2016; Liao et al., 2017).

3.2 FEATURES OF INDUSTRY 4.0

The primary characteristics (Figure 3.2) of 4IR are (1) horizontal integration (HI) by means of networked value chains that enable inter-corporation partnership, (2) vertical integration (VI) for stacked subsystems within a production facility in order to develop an adaptable and versatile manufacturing facility as well as system, and (3) end-to-end engineering integration (EEI) throughout every step of the value chain for facilitating customization of products. The HI of businesses and the VI of factories' interiors are the two foundations for EEI of processes. This is due to the fact that a product's lifecycle includes numerous phases that must be completed by various organizations. HI is the joining of beneficial linkages in order to facilitate collaboration across firms or organizations in the value chain (Foidl & Felderer, 2016; Wang et al., 2016). An assortment of more than one organization collaborates to produce outstanding products and services necessary for the growth of the business (Sindi & Roe, 2017). These firms can exchange data, financing, as well as materials with ease. As a result, novel advantages, networks, and approaches to business may arise. VI is the merging of several stacked subsystems inside an organization in order to build a flexible and adaptable manufacturing structure. The enterprise resource planning (ERP) system merges the organization's numerous informative subsystems. This will enable the implementation of a production system that is more flexible and easily reconfigurable (Weyer et al., 2015). This strategy will be accomplished because of BD management. EEI enables the production of customized products and services throughout the entire value chain (Stock & Seliger, 2016). A sequence of actions is engaged in the product-centered, value-generation approach. For example, customer requirements analysis, the creation of products (with their design and development), their manufacturing, and services.

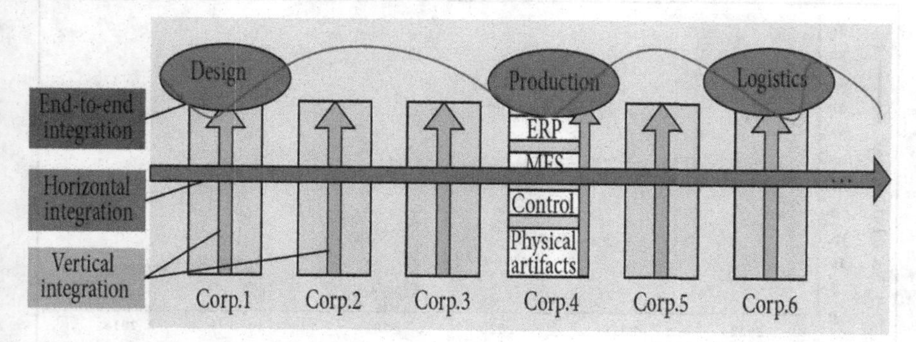

FIGURE 3.2 Integration types in manufacturing systems.

Source: Wang, S., Wan, J., Li, D., & Zhang, C. (2016). Implementing smart factory of industrie 4.0: An outlook. *International Journal of Distributed Sensor Networks*, *12*(1), 3159805. Used under CC BY-4.0.

3.3 BACKGROUND OF STUDY

A background of study is an important component of any review or research project. The appropriate literature is examined and assessed during a literature review in order to identify potential research gaps. The research gaps need to be identified and addressed in order to enhance the field of research. For this chapter, important sources of publishing on advancements in the sectors of 4IR and SM have been determined. The research publications were retrieved from the Web of Science (WoS) database by using keywords "4IR," "sustainability," and "manufacturing sector," which comprises a large number of highly esteemed publications from Emerald, Taylor and Francis, MDPI, Springer, Elsevier, and many more.

3.3.1 Year-Wise Publications

Recently, there has been a sharp increase in the number of articles published on the subject of 4IR (Figure 3.3). From 2018 to 2023, there has been a growing trend. Examining the publications by employing these keywords returned an overall total of 166 research publications in the previous six years (2018–2023) on www.webofknowledge.com. According to the data, most research was conducted in the year 2022 (55 research publications, as shown in Figure 3.3), and this number will likely continue to grow in the next years owing to the new applications and for sustaining the manufacturing sectors.

3.3.2 Country-Wise Contributions

The connections of the authors with multiple countries have been determined, revealing that Italy dominated the list, accounting for 33 of the selected 166 publications, which is almost 19.88% of contributions in total publications (as shown in Table 3.1 and Figure 3.4). India, England, and China follow with submissions

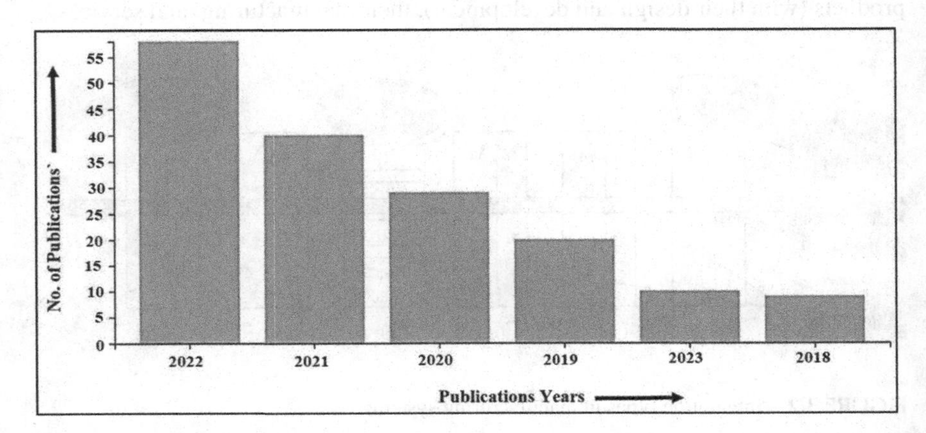

FIGURE 3.3 Detailed graphical representations of year-wise research publications (as per WoS database).

TABLE 3.1
Country-Wise Publications Details (as per WoS Database)

Sr. No.	Countries	Record Count (Out of 166 Publications)
1.	Italy	33
2.	India	27
3.	England	22
4.	China	15
5.	Malaysia	11
6.	Germany	10
7.	Portugal	9
8.	Spain	9
9.	Brazil	8
10.	France	8

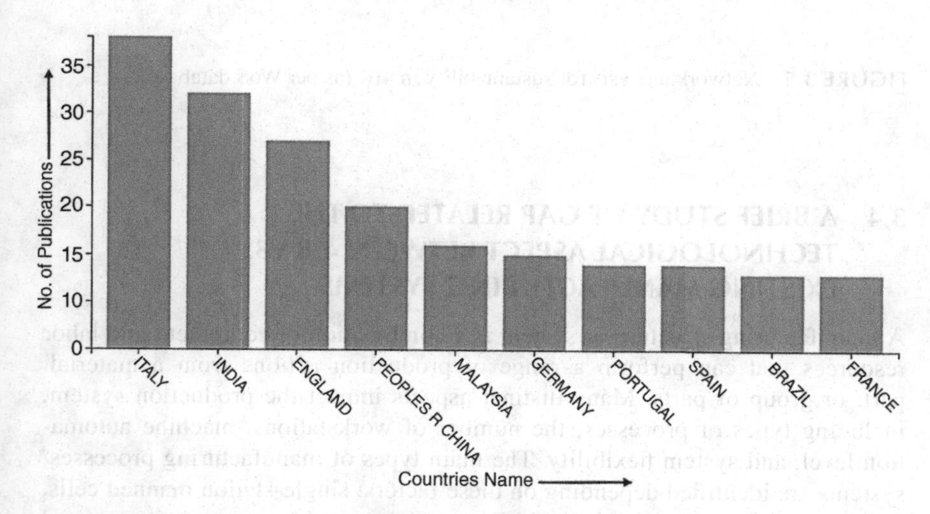

FIGURE 3.4 Country-wise analysis (≥ eight papers) (as per WoS database).

of 27, 22, and 15 papers, respectively. This demonstrates that the majority of research contributions come from Italy and Indian authors.

3.3.3 Network Analysis of Keywords

To demonstrate the link between the past studies, the VOSviewer (version 1.6.17) software was used. The researchers have employed various keywords and linked them to the concept of sustainability within the context of 4IR in their research articles. According to the results of past studies, most of the research findings have been linked with research on smart manufacturing and SM industry in connection to the manufacturing processes, additive manufacturing (AM), smart factories, sustainable industry, and so on (Figure 3.5).

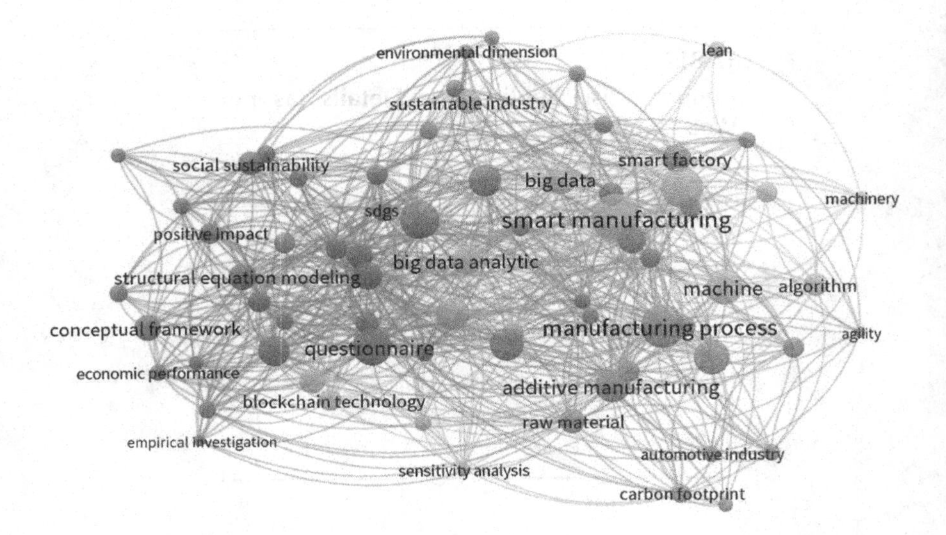

FIGURE 3.5 Network analysis for sustainability in 4IR (as per WoS database).

3.4 A BRIEF STUDY OF GAP RELATED TO THE TECHNOLOGICAL ASPECT BETWEEN 4IR VS. EXISTING MANUFACTURING SYSTEMS

A manufacturing structure or system is a combination of equipment and labor resources that can perform a range of production actions from a material, part, or group of parts. Many distinct aspects impact the production system, including types of processes, the number of workstations, machine automation level, and system flexibility. The main types of manufacturing processes/systems are identified depending on these factors: single-station manned cells, single-station automated cells (SSAC), manual assemblage system, automated assembling system (AAS), cellular manufacturing system, flexible manufacturing system (FMS), computer integrated manufacturing system (CIMS), reconfigurable manufacturing system (RMS), and many more. 4IR was conceived using this key manufacturing structure or system. (Qin et al., 2016) examined the gap between existing and future manufacturing processes, demonstrating the importance and prospects of smart manufacturing systems (SMS)/4IR innovative technologies in a manufacturing process. In comparison to smart manufacturing technologies, present manufacturing processes lack multiple elements and functionalities. Self-configuration and optimization, early awareness, persistence, and predictive and proactive maintenance are significant concerns of SMS that are currently absent in the most modern production systems, including RMS.

3.5 AM PROCESSES AND APPLICATIONS – IMPLICATIONS IN 4IR

4IR employs machinery to improve workplace conditions, including safety. It boosts output using smart technology, advanced robotics, and cloud computing. The technologies result in higher-quality products at a cheaper cost. Services, as well as product innovation, need extensive research and development, which 4IR and innovative technologies that include simulations via virtual reality provide. However, the next step involves the manufacturing processes and procedures along with their associated costs, which can impede competitiveness in the marketplace (Kim et al., 2018). Products that are one-of-a-kind may be made with no use of traditional surpluses, which is a significant benefit (Chong et al., 2018).

AM processes are organized into seven categories based on American Society of Testing and Materials (ASTM) standards. The AM operations encompass geometrical layout, the creation of computational instruments, as well as interfaces, the design of materials, process modeling, and control tools, in addition to addressing the AM application fields, which include nanoscale (bio-fabrication), micro-scale (electronics), and macro-scale (personal products, automotive), as well as large-scale (architecture and construction, aerospace and defense). The application of AM in industry includes customization, design and development, prototyping, virtual inventory, reducing wastage, and many more. "Customization" refers to the process of efficiently manufacturing a personalized and customized product utilizing various technologies. The digitization of information and customization using computer-aided manufacturing inputs are extremely speedy and efficient. Replacement of customized components consumes very minimal time and money, and it additionally enables you to meet the expectations of every customer by making new customized products on demand (Lee et al., 2014). "Design and Development": Products are swiftly created and produced using AM technologies, allowing for product enhancement and providing a superior product to the market. AM has the ability to turn imaginations into reality by introducing a product into the marketplace (Kamble et al., 2018; Long et al., 2016). "Prototyping": Before the final product goes into production, a prototype is created that can be readily manufactured utilizing a CAD file. It is capable of bringing things to market rapidly. "Reducing Wastage": This occurs due to recycling the input materials. As a result, the material managed in the form of powder is quickly recycled. Because there is less waste of raw materials, the ultimate price of the product is lower (Kolberg & Zühlke, 2015).

3.6 RECENT RESEARCH ON SUSTAINABILITY IN MANUFACTURING SECTORS

SM is a practice that considers the negative effects of industrial operations on the environment, conserving energy and natural resources, social safeguarding, and customers while being commercially competitive. In order for manufacturing holders to reap significant economic and environmental advantages from enterprises run sustainably, the concepts of SM must be carried over to the product distribution process. Table 3.2 summarizes and presents the various types of

TABLE 3.2
Review of Sustainability in Manufacturing Sectors

Manufacturing Sector/Techniques	Method Used	Detailed Description	Applications	References
Additive manufacturing (AM)-selective laser sintering process (SLS)	Empirical research	• Optimization of energy consumption for AM processes. • Two parameters were used: part orientation and slice thickness.	Enhances environmental performance on humans and living environment.	Panda et al. (2016)
Brazilian Industry – Cosmetic, thermoplastic products and utensils industry	Lean Manufacturing (LM) and Value Stream Mapping (VSM)	• Integration of sustainability models into VSM to identify manufacturing methods. • Three levels of sustainability were used from the LM perspective. • cost indicator determines economic and cost constraints.	Global and regional benchmarks were achieved using sustainability indicators.	Helleno et al. (2017)
Domestic and industrial applications of titanium oxide (TiO_2)	Life cycle assessment (LCA)	• Hydrometallurgical methods were used to reduce the effect of TiO_2 on the environment. • Alkaline and acid processes were used to minimize waste.	Used for minimizing waste generation by TiO_2 manufacturing.	Middlemas et al. (2015)
Wood-based manufacturing-rubberwood	LCA	• Effect of rubberwood-timber production on the environment is studied. • Research on carbon footprints was majorly focused. • Economic and social aspects of sustainability were analyzed.	Establishment of benchmarks of wood manufacturing on the environment.	Ratnasingam et al. (2017)
Computerized numerically controlled (CNC) machine	LCA and Analytic Hierarchy Process (AHP)	• Synthesis algorithm was used to calculate various factors, such as energy consumption, processing capability, and costs. • Sustainable design index (SDI) assessment indicators were used to determine the performance of a machine.	Used for analysis of the sustainable approach of CNC machine.	Fang et al. (2016)

sustainability in manufacturing sectors to examine recent developments in sustainability evaluation.

3.7 FUTURE IMPLICATIONS AND LIMITATIONS

In the future, AM, as a component of SMS, will enable improvements in technology in the manufacturing sector that fulfill various 4IR standards. It yields an adaptive market that readily fulfills client needs, and it is incorporated into the research and development (R&D) of new inventive goods. Just because of its capacity to customize, a consumer can get satisfaction at a low cost. Chang et al. (2018) addressed new techniques for the forthcoming generations of AM methods, which include micro/nanoscale, 3DP, bioprinting (biomaterials), along with 4D printing (a blend of AM with smart materials) for manufacturing complex three-dimensional characteristics in various materials or functionalities within high-resolution prints.

AM is not capable of manufacturing all products with the envisioned durability along accuracy. Because of its restricted material options, it is unable to satisfy all the demands of 4IR and does not optimize manufacturing in the scenario of a large-scale manufacturing system. Another downside of 4IR is the high-cost machine, which needs extensive R&D.

3.8 CONCLUSIONS

1. This book chapter highlights the SM sector in 4IR. A brief analysis of previous studies was conducted, which showed that in the year 2022, there was a maximum number of publications, and this number will likely continue to grow in the next years owing to the new application and for sustaining the manufacturing sectors.
2. As with the rest of the manufacturing industries, there is a significant gap between the existing and the implementation of 4IR, which has been stated clearly within this chapter. Through the integration of 3DP, various design and printing software, as well as processes, AM increased its contributions and involvement in 4IR. It satisfies several 4IR needs, such as customization, efficiency, and quick delivery, as well as waste reductions, and it is critical for the execution of 4IR.
3. The study evaluation approach has revealed substantial applications of AM in 4IR. It has a huge influence on 4IR to submerge and manage critical capabilities. AM and other SMS are adaptable, which opens a new market, as demonstrated in the literature review on sustainability in manufacturing sectors.

ACKNOWLEDGMENT

The authors are thankful to the University Centre of Research and Development, Chandigarh University, for providing the lab facilities and Department of Science and Technology (DST) (File No. SP/YO/2021/2514) for financial assistance.

REFERENCES

Chang, J., He, J., Mao, M., Zhou, W., Lei, Q., Li, X., Li, D., Chua, C. K., & Zhao, X. (2018). Advanced material strategies for next-generation additive manufacturing. *Materials, 11*(1), 166.

Chong, L., Ramakrishna, S., & Singh, S. (2018). A review of digital manufacturing-based hybrid additive manufacturing processes. *International Journal of Advanced Manufacturing Technology, 95*(5–8), 2281–2300.

Dilberoglu, U. M., Gharehpapagh, B., Yaman, U., & Dolen, M. (2017). The Role of Additive Manufacturing in the Era of Industry 4.0. *Procedia Manufacturing, 11*(June), 545–554.

Fang, F., Cheng, K., Ding, H., Chen, S., & Zhao, L. (2016). Sustainable design and analysis of cnc machine tools: Sustainable design index based approach and its application perspectives. *ASME 2016 11th International Manufacturing Science and Engineering Conference, MSEC 2016*, 3, 1–10.

Foidl, H., & Felderer, M. (2016). Research challenges of industry 4.0 for quality management. *Lecture Notes in Business Information Processing, 245*, 121–137.

Helleno, A. L., de Moraes, A. J. I., Simon, A. T., & Helleno, A. L. (2017). Integrating sustainability indicators and Lean Manufacturing to assess manufacturing processes: Application case studies in Brazilian industry. *Journal of Cleaner Production, 153*, 405–416.

Hermann, M., Pentek, T., & Otto, B. (2016). Design principles for Industrie 4.0 scenarios. *Proceedings of the Annual Hawaii International Conference on System Sciences, 2016-March*, 3928–3937.

Kamble, S. S., Gunasekaran, A., & Sharma, R. (2018). Analysis of the driving and dependence power of barriers to adopt industry 4.0 in Indian manufacturing industry. *Computers in Industry, 101*(March), 107–119.

Kim, H., Lin, Y., & Tseng, T.-L. B. (2018). A review on quality control in additive manufacturing. *Rapid Prototyping Journal, 10*(3), 157.

Kolberg, D., & Zühlke, D. (2015). Lean Automation enabled by Industry 4.0 Technologies. *IFAC-PapersOnLine, 28*(3), 1870–1875.

Lee, J., Kao, H. A., & Yang, S. (2014). Service innovation and smart analytics for Industry 4.0 and big data environment. *Procedia CIRP, 16*, 3–8.

Liao, Y., Deschamps, F., de Loures, E. F. R., & Ramos, L. F. P. (2017). Past, present and future of Industry 4.0 – A systematic literature review and research agenda proposal. *International Journal of Production Research, 55*(12), 3609–3629.

Liu, Y., & Xu, X. (2017). Industry 4.0 and cloud manufacturing: A comparative analysis. *Journal of Manufacturing Science and Engineering, Transactions of the ASME, 139*(3), 034701.

Long, F., Zeiler, P., & Bertsche, B. (2016). Modelling the production systems in industry 4.0 and their availability with high-level Petri nets. *IFAC-PapersOnLine, 49*(12), 145–150.

Luthra, S., Govindan, K., Kannan, D., Mangla, S. K., & Garg, C. P. (2017). An integrated framework for sustainable supplier selection and evaluation in supply chains. *Journal of Cleaner Production, 140*, 1686–1698.

Mangla, S. K., Kumar, P., & Barua, M. K. (2015). Risk analysis in green supply chain using fuzzy AHP approach: A case study. *Resources, Conservation and Recycling, 104*, 375–390.

Middlemas, S., Fang, Z. Z., & Fan, P. (2015). Life cycle assessment comparison of emerging and traditional Titanium dioxide manufacturing processes. *Journal of Cleaner Production, 89*, 137–147.

Panda, B. N., Garg, A., & Shankhwar, K. (2016). Empirical investigation of environmental characteristic of 3-D additive manufacturing process based on slice thickness and part orientation. *Measurement: Journal of the International Measurement Confederation, 86*, 293–300.

Qin, J., Liu, Y., & Grosvenor, R. (2016). A categorical framework of manufacturing for Industry 4.0 and beyond. *Procedia CIRP, 52*, 173–178.

Ratnasingam, J., Ramasamy, G., Ioras, F., & Parasuraman, N. (2017). Assessment of the Carbon Footprint of Rubberwood Sawmilling in Peninsular Malaysia: Challenging the Green Label of the Material. *BioResources, 12*(2), 3490–3503.

Sarkis, J., & Zhu, Q. (2018). Environmental sustainability and production: Taking the road less travelled. *International Journal of Production Research, 56*(1–2), 743–759.

Sindi, S., & Roe, M. (2017). Strategic supply chain management: The development of a diagnostic model. *Strategic Supply Chain Management: The Development of a Diagnostic Model*, 1–272.

Stock, T., & Seliger, G. (2016). Opportunities of sustainable manufacturing in Industry 4.0. *Procedia CIRP, 40*(Icc), 536–541.

Thompson, M. K., Moroni, G., Vaneker, T., Fadel, G., Campbell, R. I., Gibson, I., Bernard, A., Schulz, J., Graf, P., Ahuja, B., & Martina, F. (2016). Design for additive manufacturing: Trends, opportunities, considerations, and constraints. *CIRP Annals – Manufacturing Technology, 65*(2), 737–760.

Tjahjono, B., Esplugues, C., Ares, E., & Pelaez, G. (2017). What does Industry 4.0 mean to Supply Chain? *Procedia Manufacturing, 13*, 1175–1182.

Wang, L., & Wang, G. (2016). Big data in cyber-physical systems, digital manufacturing and Industry 4.0. *International Journal of Engineering and Manufacturing, 6*(4), 1–8.

Wang, S., Wan, J., Li, D., & Zhang, C. (2016). Implementing smart factory of Industrie 4.0: An Outlook. *International Journal of Distributed Sensor Networks, 2016.*

Weyer, S., Schmitt, M., Ohmer, M., & Gorecky, D. (2015). Towards industry 4.0 – Standardization as the crucial challenge for highly modular, multi-vendor production systems. *IFAC-PapersOnLine, 28*(3), 579–584.

4 Plastic Waste Recycling by Additive Manufacturing

Sustainable Approach to Circular Economy

Khushwant Kour, Ketan Badogu, and Ranvijay Kumar
Chandigarh University, Mohali, India

4.1 INTRODUCTION

The chemical and mechanical qualities of plastic products are diverse, making them suitable for a broad spectrum of applications. Globally, plastic manufacturing is increasing, and 300 million metric tons of plastic garbage are produced each year. However, the problem of the nondegradation property of plastic impacts the environment, and plastic waste contamination poses a serious concern (Ryberg et al., 2019). The development of a circular economy has been a top approach for scholars in the realm of sustainable development of plastic waste in recent years (Sauerwein et al., 2019). A rapidly expanding group of production methods known as additive manufacturing (AM), commonly referred to as 3D printing, enables novel manufacturing perspectives. The literature has been paying more attention to how AM contributes to sustainability (Diegel et al., 2016). Based on the prospects of AM for reducing resource consumption, it is thought to be viable for sustainable manufacturing. In comparison with subtractive processes like milling, the addictive nature of AM prevents material losses, and for developing solutions for the circular economy, these factors could potentially present innovative possibilities for the reduction of plastic waste (Mani et al., 2014).

Products made from polymers and recycled materials can be produced using AM. In addition, previous studies have outlined that the laser-based AM is anticipated to be essential for circular economy, especially for metal repair (Leino et al., 2016). The mechanical characteristics, forms, sizes, and shades of the polymers employed in diverse goods vary, creating obstacles to their efficient disposal. To effectively manage polymer waste, polymers must be treated and separated based

DOI: 10.1201/9781003309123-4

on their chemical and physical characteristics. Recycling is one technique used to refurbish waste polymers into new beneficial goods. However, it is a complicated procedure that requires several phases of processing. Nevertheless, reusing and recycling can extend the life of polymers. In this manner, AM may catalyze the development of new products using waste plastics and the implementation of a recycling and reusing framework at the local scale (Colorado et al., 2020).

The environmental effect of an improvement is currently one of the most important factors in determining whether it is practical and sustainable. This study provides a thorough analysis of the sustainability of AM, including everything from circular economy and material recycling to other environmental issues. It also raises awareness about the potential and implications of using the AM, as well as promotes sustainability by emphasizing the importance of taking the appropriate steps to comply with the reuse and recycling of plastic waste.

4.2 BACKGROUND OF THE STUDY

The background of the study has been described following the information obtained from www.webofknowledge.com (as shown in Figure 4.1, Tables 4.1 and 4.2). Examining the research papers from the last five (2017–2022) years using the keywords "plastic waste recycling" and "AM" also produced a total of 158 publications. The largest number of studies were reported statistically in 2022, and this number is anticipated to increase in the next years owing to fresh applications (see Figure 4.1). Table 4.1 shows the analyzed data table of "Plastic waste recycling" and "AM" of publishing years 2017–2022.

The VOSviewer software program (version 1.6.18) was used to identify keywords of common topics of plastic waste recycling research in the context of construction based on the co-occurrence of terms. Figure 4.2 displays the map

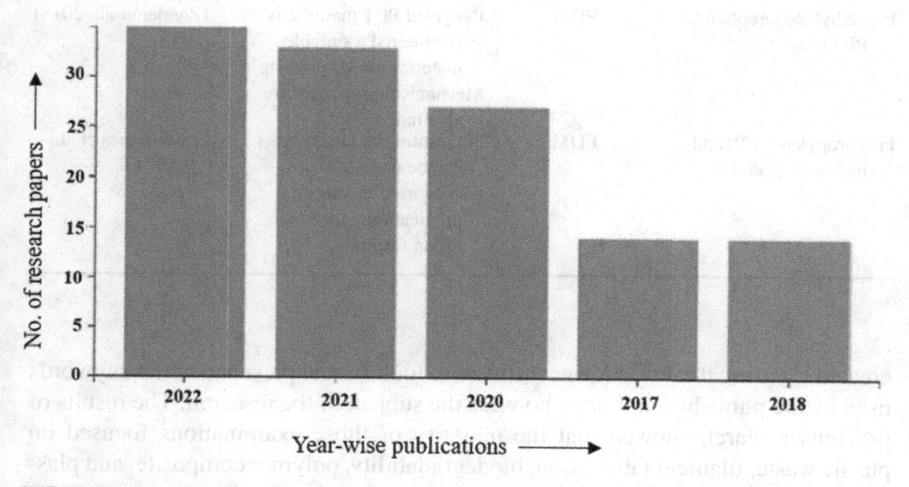

FIGURE 4.1 Year-wise publications in plastic waste recycling and AM field.

TABLE 4.1
Analyzed Data Table of "Plastic Waste Recycling" and "AM" of Publishing Years (2017–2022) (as per web of science database)

Publication Years	Record Count	% of 158
2022	35	22.152
2021	33	20.886
2020	27	17.089
2017	14	8.861
2018	14	8.861

TABLE 4.2
Applications of the Recyclable Materials and Their Outputs

Materials and Reinforcements	3D Printing Technique	Output/Remarks	References
Polylactic acid (PLA) and carbon fiber (CF)	FDM	Recyclable composites can be used for low-cost industrial applications.	(Tian et al., 2016)
High-density polyethylene (HDPE), silicon carbide/aluminum oxide (SiC/Al2O3)	FDM	Reinforced composite has better wear properties. Can be used for rapid tooling applications.	(Singh et al., 2018)
Polyamide 12 (PA-12), TPU, aramid, and graphite	FDM	Composite material blends show a 1.5 time increase in modulus, as well as elongation at the break in comparison with pure PA.	(Mägi et al., 2016)
Polyethylene terephthalate (PET)	FDM	Recycled PET material is considered a suitable material for 3D printing. Mechanical properties are also tuned.	(Zander et al., 2018)
Polypropylene (PP) and tire waste granulate	FDM	3D printing of large parts can be done. Can be used in various applications, such as urban furniture.	(Domingues et al., 2017)

created by using the VOSviewer software to look for the presence of the keywords used by the published authors who were the subject of the research. The results of previous research showed that the majority of those examinations focused on plastic waste, filament fabrication, biodegradability, polymer composite, and plastic recycling using AM.

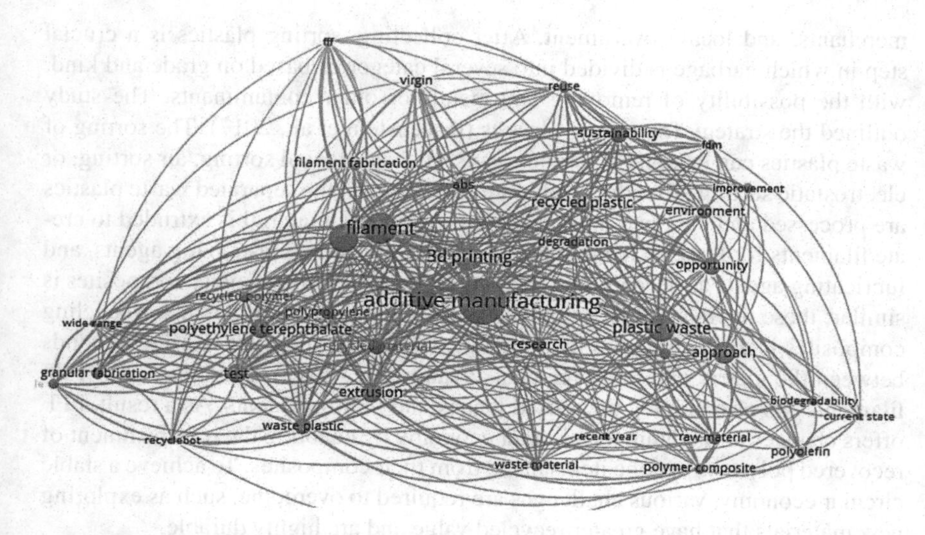

FIGURE 4.2 The relation between the terms reported in research papers on plastic waste recycling and AM.

4.3 PLASTIC WASTE RECYCLING IN AM

To reduce waste production and adverse environmental effects in all industrial processes, various efforts should be undertaken. When determining the environmental effect, major factors are included, such as the type of initial supplies employed in the creation process and how the waste is disposed of. Among these AM methods, fused filament fabrication (FFF) is a well-liked method for producing polymers and their composites (Herianto & Mastrisiswadi, 2020). In the FFF method, recyclable thermoplastic materials may be employed as raw materials. All residual thermoplastic material or waste plastic can be recycled and utilized in AM.

In general, there are three basic techniques for plastic waste recycling: mechanical recycling (MR), incineration or burning, and chemical recycling (CR) (Bartolome et al., 2012). However, if the characteristics are optimized and the printing requirements are achieved, recycled polymers from MR and CR techniques may be utilized to produce AM feedstock (Rahimizadeh et al., 2019). In the MR method, the polymer is purified from impurities, crushed into granulates, and then released to create the required component. During CR, the polymer chains are severed, either completely returning to monomers or partially returning to oligomers, and the incineration or burning process is not an environmentally friendly recycling technique, but it does recover the chemical energy trapped in the polymers as heat energy. The previous study has outlined the recycling of polymeric materials for AM (Zander, 2019).

The initial stage in the recycling process is the collection of plastic garbage. Direct collection of plastic garbage can be obtained from businesses, large merchants, and local government. After collecting, sorting plastics is a crucial step in which garbage is divided into several categories based on grade and kind, with the possibility of removing any metals or other contaminants. The study outlined the strategy for sorting plastics (Rozenstein et al., 2017). The sorting of waste plastics can be done in various ways, such as manual sorting, air sorting, or electrostatic sorting. To remove any foreign elements, the separated waste plastics are processed and chopped into bits. After that, waste material is extruded to create filaments for the FFF technique by adding stabilizers, plasticizing agents, and lubricating agents as needed. The separation of fibers from obsolete composites is similar; these fibers are then employed as reinforcements in filaments. Recycling composites is difficult due to their chemical diversity and tight interfacial bonds between the matrix and reinforcements (Yang et al., 2012). The creation of FFF filaments can also employ the reused matrix and reinforcements. As a result, FFF offers chances to maintain the circular economy by enabling the refurbishment of recovered polymers and the items made from their composites. To achieve a stable circular economy, various challenges are required to overcome, such as exploring new materials that have greater recycled value and are highly durable.

4.4 RECYCLING POLYMER MATERIALS AND THEIR COMPOSITES USING 3D PRINTING TECHNIQUE

By using recycled and reused polymers and plastic waste in AM, the waste output will be decreased, and their value will be increased. This could serve as a way to recycle unwanted polymers for use in contemporary applications. Yet there hasn't been much study done in this area, and it could be possible to improve the value of fused deposition modeling (FDM) components by creating reinforced recycled polymers. Table 4.2 depicts the properties and characteristics of recycling materials and their composites.

4.5 FUTURE PERSPECTIVE AND ENVIRONMENTAL EFFECTS OF RECYCLED WASTE PLASTICS

Compared to traditional machining, 3D printing offers an aggregate decrease in environmental burden, notably in terms of both energy and material usage because it significantly reduces the amount of trash created throughout the recycling process (Mami et al., 2017).

The analysis of recycled plastic resources into raw materials for AM still faces various obstacles. One of the primary obstacles to adopting recycled materials is the absence of standardization and quality assurance. The variety of mechanical properties and limits of recycled materials that can be replaced need to be clarified. It would also be beneficial to reduce the number of stages required to recycle plastic waste into raw materials. However, some raw materials and recycled plastic materials used in AM can produce health risks in the form of organic chemicals, including styrene, butane, ethylbenzene, etc. In this regard, the temperature of the

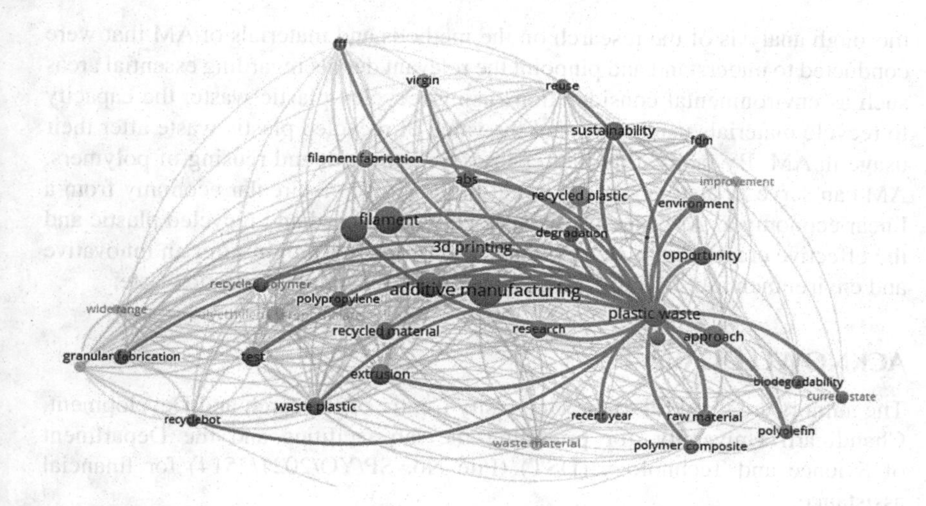

FIGURE 4.3 Gap in previous studies related to plastic waste recycling and AM.

extruder and printer failure both have a significant impact on particle emissions. The negative health effects of 3D printing are related to chemical gas emissions, which in places with adequate ventilation seem to be minimal. When comparing 3D printers utilizing acrylonitrile butadiene styrene (ABS) to those using thermoplastic polylactcic acid (PLA) material, the overall ultrafine particle (UFP) is around an order of magnitude higher for the ABS material. Yet, both of these materials have the potential to be classified as "high emitting materials" of UFP, so caution must be exercised when using such filaments in enclosed spaces without proper ventilation or filtration (Stephens et al., 2013). Future research is, therefore, necessary to create less hazardous materials from recycled waste materials and establish guidelines for manufacturing safety. Furthermore, the use of recycled materials may increase as a result of an open-source knowledge-sharing resource that covers all topics, including filament manufacturing, FFF build settings, and material attributes. Figure 4.3 shows the gaps in reported terms for previous studies in relation to Figure 4.2. As depicted in Figure 4.3, the improvement and tuning of the mechanical, thermal, and chemical properties may be accomplished by conducting future studies related to plastic recycling through AM. The previous studies have been less reported on the biodegradability of the additively manufactured parts prepared from plastics. Future studies may also be conducted for detailed biodegradability analysis of additively manufactured plastics.

4.6 CONCLUSION

Despite the many benefits that plastic materials offer to society and the economy, the environment is severely burdened by its inability to degrade because there is no systematic method for processing plastic waste. This chapter provides a thorough analysis of the research on the methods and materials of AM that were conducted to understand and pinpoint the relevant details regarding essential areas such as environmental considerations using recycled plastic waste, the capacity to recycle materials using AM, and the value of recycled plastic waste after their usage in AM. By facilitating the repair, refurbishment, and reusing of polymers, AM can serve as a strategic tool to aid in the shift to a circular economy from a linear economy as AM fills the distance between reused and recycled plastic and the effective creation of a new product. Therefore, AM might offer an innovative and environmentally friendly perspective on the recycling of plastic waste.

ACKNOWLEDGMENT

The authors are thankful to the University Centre of Research and Development, Chandigarh University, for providing the lab facilities and the Department of Science and Technology (DST) (File No. SP/YO/2021/2514) for financial assistance.

REFERENCES

Bartolome, L., Imran, M., Gyoo, B., & Hyun, D. (2012). Recent Developments in the Chemical Recycling of PET. *Material Recycling – Trends and Perspectives*. https://doi.org/10.5772/33800

Colorado, H. A., Velásquez, E. I. G., & Monteiro, S. N. (2020). Sustainability of additive manufacturing: the circular economy of materials and environmental perspectives. *Journal of Materials Research and Technology*, 9(4), 8221–8234. https://doi.org/10.1016/j.jmrt.2020.04.062

Diegel, O., Kristav, P., Motte, D., & Kianian, B. (2016). *Handbook of Sustainability in Additive Manufacturing*, 73–99. https://doi.org/10.1007/978-981-10-0606-7

Domingues, J., Marques, T., Mateus, A., Carreira, P., & Malça, C. (2017). An additive manufacturing solution to produce big green parts from tires and recycled plastics. *Procedia Manufacturing*, 12(December 2016), 242–248. https://doi.org/10.1016/j.promfg.2017.08.028

Herianto, Atsani S. I., & Mastrisiswadi, H. (2020). Recycled polypropylene filament for 3D printer: extrusion process parameter optimization. *IOP Conference Series: Materials Science and Engineering*, 722(1). https://doi.org/10.1088/1757-899X/722/1/012022

Leino, M., Pekkarinen, J., & Soukka, R. (2016). The role of laser additive manufacturing methods of metals in repair, refurbishment and remanufacturing – Enabling circular economy. *Physics Procedia*, 83, 752–760. https://doi.org/10.1016/j.phpro.2016.08.077

Mägi, P., Krumme, A., & Pohlak, M. (2016). Recycling of PA-12 in additive manufacturing and the improvement of its mechanical properties. *Key Engineering Materials*, 674, 9–14. https://doi.org/10.4028/www.scientific.net/KEM.674.9

Mami, F., Revéret, J. P., Fallaha, S., & Margni, M. (2017). Evaluating eco-efficiency of 3D printing in the aeronautic industry. *Journal of Industrial Ecology*, 21, S37–S48. https://doi.org/10.1111/jiec.12693

Mani, M., Lyons, K. W., & Gupta, S. K. (2014). Sustainability characterization for additive manufacturing. *Journal of Research of the National Institute of Standards and Technology*, 119, 419–428. https://doi.org/10.6028/jres.119.016

Rahimizadeh, A., Kalman, J., Fayazbakhsh, K., & Lessard, L. (2019). Recycling of fiberglass wind turbine blades into reinforced filaments for use in Additive Manufacturing. *Composites Part B: Engineering*, 175, 107101. https://doi.org/10.1016/j.compositesb.2019.107101

Rozenstein, O., Puckrin, E., & Adamowski, J. (2017). Development of a new approach based on midwave infrared spectroscopy for post-consumer black plastic waste sorting in the recycling industry. *Waste Management*, 68, 38–44. https://doi.org/10.1016/j.wasman.2017.07.023

Ryberg, M. W., Hauschild, M. Z., Wang, F., Averous-Monnery, S., & Laurent, A. (2019). Global environmental losses of plastics across their value chains. *Resources, Conservation and Recycling*, 151(August), 104459. https://doi.org/10.1016/j.resconrec.2019.104459

Sauerwein, M., Doubrovski, E., Balkenende, R., & Bakker, C. (2019). Exploring the potential of additive manufacturing for product design in a circular economy. *Journal of Cleaner Production*, 226, 1138–1149. https://doi.org/10.1016/j.jclepro.2019.04.108

Singh, N., Singh, R., & Ahuja, I. P. S. (2018). Recycling of polymer waste with SiC/Al2O3 reinforcement for rapid tooling applications. *Materials Today Communications*, 15, 124–127. https://doi.org/10.1016/j.mtcomm.2018.02.008

Stephens, B., Azimi, P., El Orch, Z., & Ramos, T. (2013). Ultrafine particle emissions from desktop 3D printers. *Atmospheric Environment*, 79, 334–339. https://doi.org/10.1016/j.atmosenv.2013.06.050

Tian, X., Liu, T., Wang, Q., Dilmurat, A., Li, D., & Ziegmann, G. (2016). Recycling and remanufacturing of 3D printed continuous carbon fiber reinforced PLA composites. *Journal of Cleaner Production*, 1–10. https://doi.org/10.1016/j.jclepro.2016.11.139

Yang, Y., Boom, R., Irion, B., van Heerden, D. J., Kuiper, P., & de Wit, H. (2012). Recycling of composite materials. *Chemical Engineering and Processing: Process Intensification*, 51, 53–68. https://doi.org/10.1016/j.cep.2011.09.007

Zander, N. E. (2019). Recycled polymer feedstocks for material extrusion additive manufacturing. In *Polymer-based additive manufacturing: recent developments, American Chemical Society. Chapter 3, ACS Symposium Series* Vol. 1315 (pp. 37–51).

Zander, N. E., Gillan, M., & Lambeth, R. H. (2018). Recycled polyethylene terephthalate as a new FFF feedstock material. *Additive Manufacturing*, 21, 174–182. https://doi.org/10.1016/j.addma.2018.03.007

5 Zero-Waste Manufacturing
Current Scenario

Harpreet Kaur Channi and Pulkit Kumar
Chandigarh University, Mohali, India

5.1 INTRODUCTION

The creation of solid waste in the globe has been substantially accelerated, particularly in nations that are still in the process of economic development, urbanization, and rising community living standards (Turok and McGranahan 2013). Increasing population, booming economy, and fast urbanization are all factors that have contributed to this acceleration (Nguea 2023). The disposal of solid waste is a major issue for environmental sustainability in the modern world (Naveenkumar et al. 2023). Misallocated resources are represented as waste, which is a symbol of the inefficiency that affects modern civilizations. It is estimated that 11 billion tons of solid trash are produced annually (enough to drive 2.5-ton automobiles 300 times around the equator), with each individual producing around 1.74 tons of solid waste every year (Vinti et al. 2023). The strong demand for new products, however, results in the daily depletion of a vast number of natural resources, in addition to the massive amounts of garbage created. Because of the high demand for innovative goods, this is the case (Boadway and Flatters 2023). The yearly consumption of 120–130 billion tons of natural resources produces 3.4–4 billion tons of municipal solid garbage. Any waste generated requires more money to deal with since it consumes more resources, creates more work for workers, puts more stress on the land, pollutes the environment, and so on. Authorities are under intense pressure to find more sustainable solutions for rubbish disposal as the volume of trash continues to grow (Farooqi et al. 2021).

As civilization transitions from rural to urban, solid waste management becomes crucial. Industrialization has brought many substances that nature cannot or slowly digest. Due to their restricted degradability or toxicity, certain industrial products include substances that may build up in nature and threaten humanity's future use of natural resources like drinking water, agricultural land, air, etc. City planners have paid less attention to waste management than they have to utilities like water and power (Aqilah et al. 2023). Waste management is an area where current planning is lacking (Burke et al. 2023). The repercussions

 DOI: 10.1201/9781003309123-5

of global warming on today's civilization have made it more eco-friendly. Reduced global resources also increase the need for responsible resource and product management. As societies transition from rural areas with low population densities to urban areas with dense populations, the management of solid waste takes on more significance. As a byproduct of industrialization, we now have many products that nature cannot process or processes very slowly (Sohail et al. 2023).

Recent years have seen the rise of the zero-waste movement. Zero waste aims to recycle everything. No landfills or incinerators receive waste. We encourage recycling natural materials. In 1973, Dr. Paul Palmer coined "zero waste" to explain chemical reuse. Every usable item is reused in a waste-free method. Nothing is wasted in a circular system. This results in goods being reused, resold, or mended (Weber, et al. 2023). Instead of exploitation, natural resources may be recycled or recovered from trash and used as inputs. The goal of zero-waste initiatives is to eliminate the need for garbage collection and disposal. Lessen your waste by reusing and recycling as much as you can. Though the goal of zero waste may seem impossible, even the smallest contribution may make a difference. Avoiding single-use items and recycling are great ways to cut down on trash (Gale 2023). Together, everyone's efforts can make a big difference, and even the smallest changes may have a positive impact. The zero-waste movement promotes fixing, reusing, and recycling as alternatives to throwing things away. The zero-waste philosophy is an approach to production and consumption that aims to minimize garbage (Ali and Shirazi 2023). By doing so, we may reduce our impact while also reducing methane emissions and protecting the environment (Wikurendra et al. 2023). The goal of the zero-waste movement is to lead a waste-free existence. We can get closer to our goal if we start paying more attention to our purchasing habits. Sustainable packaging informs consumers about the item's impact on the planet. The eco-friendliness of a product may be verified by reading the label. The term "zero waste" may refer to both the manufacturing process and the final product (Tan et al. 2023).

Manufacturers are using more sustainable methods due to the rising demand for zero-waste products and manufacturing. Companies that use trash as raw material might contribute to a waste-free society. A zero-waste system recycles or reuses all rubbish instead of sending it to a landfill. Under the current production system, it is the customer who throws away unwanted goods. Waste management as a "whole system" is prioritized by the zero-waste concept above individual item disposal (Tan et al. 2023).

Product stewardship, also known as Extended Producer Responsibility, is an initiative that urges product manufacturers, packagers, and disposers to shoulder more of the burden associated with their products' lifetimes. Recycled materials should be used wherever possible, and manufacturers should create long-lasting products that can be recycled and reused. Zero waste means there is no waste at all (Makinde et al. 2023). Waste is avoided, and all materials are put to good use in a zero-waste system. The disposal of trash in landfills, dumps, or incinerators is outlawed. Zero-waste practices concentrate on minimizing waste. It involves making efficient use of resources and reusing or recycling materials for the sake of human and environmental health. The Zero Waste International Alliance is an organization dedicated to reducing waste throughout the world by encouraging recycling, reusing, and composting (Makinde et al. 2023).

Poor garbage disposal, open dumping, and landfilling threaten air, land, and water environments and health. Landfill leachate pollutes ground and surface water, while carbon monoxide and methane from rubbish heaps contaminate the air. Waste management that emphasizes prevention, reduction, remediation, and reuse has the potential to save resources and keep the environment safe (Debnath et al. 2023). Production systems that consume few inputs, generate little waste and recycle or reuse their by-products are at the heart of the zero-waste manufacturing (ZWM) approach. Waste prevention, waste reduction, recycling, product redesign, and creative reuse are all part of its toolkit. Reduced manufacturing waste might result from techniques that extend a product's useful life beyond its initial intended use (Esmaeilian et al. 2018). This means that in order to achieve zero waste, goods must have many purposes and a long, dependable lifespan across a variety of applications. The normal manufacturing process challenges sustainable material use, lean material generation, and minimal material removal. ZWM necessitates the development of novel production techniques (Kumari and Raghubanshi 2023). Figure 5.1 shows the benefits of ZWM.

The pace of technology and the changing climate characterize our era. Big data, the Internet of Things, cloud computing, cognitive processing systems, and artificial intelligence (AI) are just a few examples of digital technologies that are experiencing technology acceleration or exponential growth. The growth of these

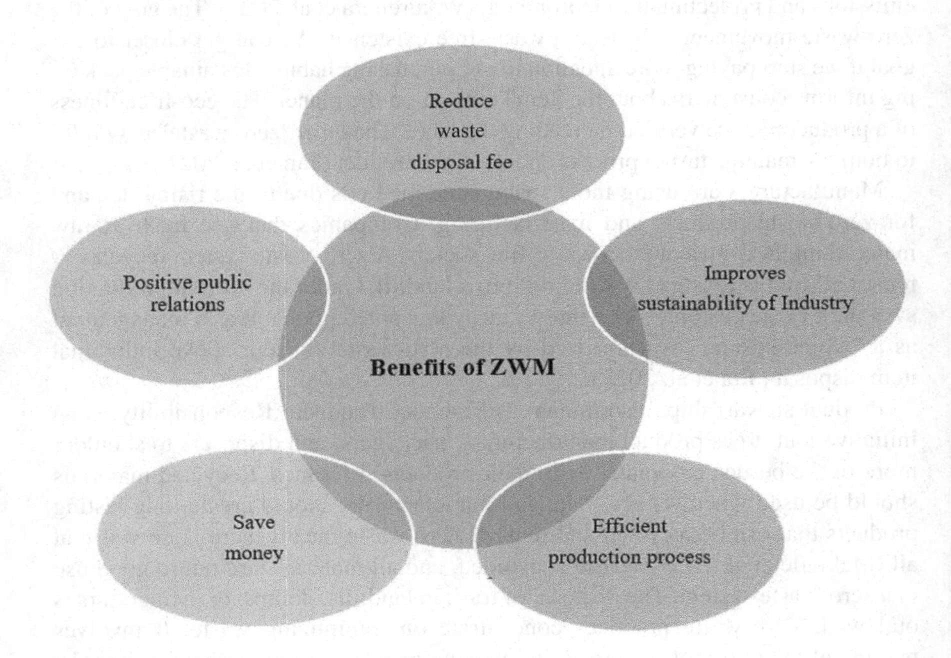

FIGURE 5.1 Benefits of ZWM.

technologies is reflected in statistics like the yearly growth rate of data storage capacity and the number of devices with internet access. The rapid development of technology may disrupt business as usual, but it also opens up exciting new avenues for growth. It is one of the main forces behind Industry 4.0, which is causing a revolution in production methods that rely on Cyber-Physical Production Systems (CPPS). The second option helps businesses save time and money by enhancing efficiency and productivity in the industrial sector (Pozzi et al. 2023). As a result of increasing levels of greenhouse gases in the atmosphere, global weather patterns are shifting. Both businesses and governments have made it a top priority to strive toward a more responsible reduction of CO_2 emissions in response to the growing threat of climate change. In a recent announcement, the European Commission unveiled the European Green Deal. The latter refers to plans to eliminate all carbon emissions in Europe by the year 2050. Additionally, industrial corporations are adopting environmentally friendly practices. Climate change and the rapid development of technology are interconnected. The digital transformation of industrial organizations may lead to lower CO_2 emissions and greater sustainability via more eco-friendly product design, more efficient waste management, and more recycling and other circular economy practices (Guandalini 2022).

This is why the twin transition is so important to industrial enterprises and governments. According to the latter, businesses need to undergo simultaneous digital and environmental reforms. The end goal is to aid in the transition to a carbon-free, waste-free, resilient, and inclusive economy. Twofold transformation in industrial organizations is driven by Industry 4.0. Industry 4.0's use of predictive maintenance, digital quality management, and zero-defect manufacturing (ZDM) increases product quality and decreases waste (Psarommatis et al. 2020). Stakeholders may cut carbon dioxide (CO_2) emissions and embrace circular economy services with the use of Industry 4.0 supply chains. Quality control in manufacturing might be enhanced through CPPS systems and digital technologies. Failure and defect detection in a production line may be enhanced by the collection and analysis of massive amounts of digital data. They help factories anticipate and avoid issues as well. Predictive quality is a widely appreciated use of quality management. Combining proactive and reactive strategies from maintenance, logistics, and process control may help ensure that only defect-free products leave the facility. ZDM's mission guarantees high standards. After a decade, the cost and feasibility of ZDM are greatly improved by Industry 4.0. Quality management and ZDM are very helpful for manufacturing organizations. They also help the environment in other ways. Defect detection helps cut down on waste and quality issues. Waste is reduced when defective products are avoided via data-driven process management. When it comes to servicing and fixing assets, predictive maintenance is the way to go. The result is better utilization of assets and less waste and garbage due to malfunctions. Asset, process, and plant levels are all suitable for implementing quality management and ZDM strategies. Sustainability may be enhanced by using closed-loop control and multilevel methods. The latter finds the problem and fixes it. Improved economic and environmental sustainability in manufacturing is a result of using ZDM and quality management systems (Powell et al. 2022).

The scope of a typical ZDM strategy might range from a single building to the whole infrastructure that supports it. Additional opportunities for recycling waste and meeting supply-chain-wide sustainability goals are made possible by the value chains that characterize the circular economy (CE). The inefficient "take-make-dispose" model has been updated in the modern CE era. The CE's value chains make available restorative and regenerative practices that may increase sustainability. When one company uses another's goods or trash, it is necessary to remanufacture, reuse, and repurpose the materials. Repairing and reusing a car battery, for example, is an option if its individual cells still have some life left in them. The possibilities of the CE value chain are enhanced by Industry 4.0 because of the digitization of product data and expertise and the easy communication between manufacturers and remanufacturers. Providing traceability of materials and visibility into the circular chain increases the openness and acceptance of the CE process by the end users. Industry 4.0 applications may use CPPS systems to keep everyone in the recycling and waste disposal cyclical chain updated (Pérez et al. 2015). The digitization of manufacturing processes allows for the adoption of greener methods like the CE. Industry 4.0 facilitates the comprehensive and standardized greening of manufacturing businesses. Key performance indicators (KPIs), including carbon dioxide (CO_2) emissions, garbage, and recycled waste, are calculated to evaluate the company's environmental performance. The computation and monitoring of such KPIs may aid in the implementation of sustainability plans at multiple levels, including individual assets, industrial processes, factories, and manufacturing organizations. Adopting digital manufacturing systems that allow for the definition, computation, and monitoring of key performance metrics is a smart move for businesses in the manufacturing industry.

Zero-waste production was an emerging idea that sought to reduce waste throughout the manufacturing process. It entailed creating goods and methods to reduce waste by lowering resource use, recycling materials, and reusing components. ZWM may have improved as the manufacturing business is always changing (Pérez et al. 2015). However, various projects and technologies were being deployed to achieve ZWM, as shown in Figure 5.2:

- **Design for Disassembly**: Designers were putting a lot of effort into producing items that could be disassembled and recycled. Only the parts that are no longer needed need to be disposed of, rather than the whole product.
- **Closed-Loop Systems**: Many companies now use "closed-loop" systems, in which both inputs and outputs (waste) remain inside the same cycle. Thus, less energy is used, and less waste is generated.
- **Lean Manufacturing**: In order to increase productivity and decrease waste, several companies adopted lean manufacturing practices. Wasteful procedures were identified and eliminated with the use of value stream mapping, continuous improvement, and just-in-time production.
- **Material and Energy Efficiency**: There was a shift toward using more energy- and resource-efficient materials and methods of production throughout businesses. These initiatives included using renewable energy sources,

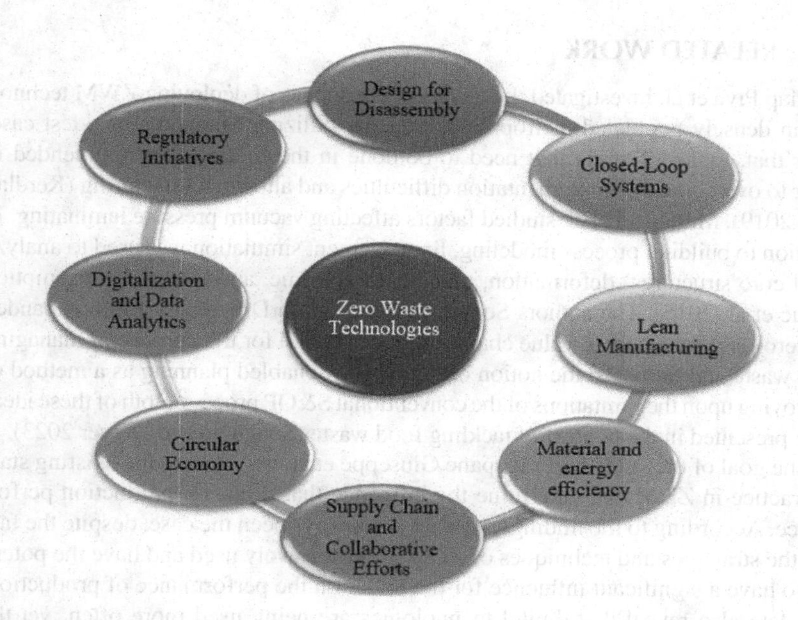

FIGURE 5.2 Zero-waste technologies.

improving energy efficiency, and decreasing dependency on potentially dangerous compounds. There were certain industries that had waste-to-energy systems set up, which used rubbish as fuel. This aids in the generation of renewable energy and the prevention of waste from going to landfills.

- **Supply Chain and Collaborative Efforts**: Companies were urged to coordinate with their suppliers and local waste management organizations to identify further avenues for trash reduction. To achieve zero-waste goals as a community, it was necessary to share knowledge, methods, and materials. It is crucial to learn from already existent examples of collaboration between producers, distributors, and end users in order to increase sustainability throughout the whole supply chain as a whole.
- **Circular Economy**: Analyze the impact that the CE, which aims to maximize resource utilization via reuse, refurbishment, and recycling, has on zero-waste production methods. Cutting-edge recycling technologies, such as chemical recycling and enhanced sorting techniques, should be considered for better management of industrial waste.
- **Digitalization and Data Analytics**: Think about how the industrial industry is using digital technologies, data analytics, and AI to boost efficiency, reduce waste, and save money. Efforts to Regulate: Keep abreast of any policy changes or legislation that may affect zero-waste practices, particularly as they pertain to waste management and sustainability in manufacturing (Meng et al. 2023).

5.2 RELATED WORK

Kerdlap Piya et al. investigated the technical challenges of deploying ZWM technology in densely populated metropolitan areas by utilizing Singapore as a test case. After that, topics of study that need to be done in the future are recommended in order to overcome the implementation difficulties and allow manufacturing (Kerdlap et al. 2019). Mario Lui et al. studied factors affecting vacuum pressure laminating. In addition to building process modeling, finite element simulation was used to analyze mold core structures' deformation, manufacturing time, and material consumption (Lušić et al. 2015). The authors Sengupta, Sourav, and Dreyertudy Heidi expanded the zero-defect concept to value chains in grocery retail for the purpose of managing food waste and proposed the notion of digital twin-enabled planning as a method of improving upon the limitations of the conventional S&OP process. Both of these ideas were presented in the context of tackling food waste (Sengupta and Dreyer 2023).

The goal of the study by Fragapane Giuseppe et al. was to map the existing state of practice in ZDM and determine the influence that it has on production performance. According to the findings, this has not always been the case, despite the fact that the strategies and techniques of ZDM are extensively used and have the potential to have a significant influence for the better on the performance of production. The data also reveal that digital technologies are being used more often, yet the potential of AI and extended reality is still being underutilized (Fragapane et al. 2023). New product manufacturing restrictions by Psarommatis Foivos and Bravos George et al. reduced production time and batch size. Previously successful methods become useless or antiquated. To embrace agility and provide a wide range of services to the industrial sector, digitally connected industrial automation will challenge digital boundaries across industrial systems. Chen Jiaxin and colleagues developed a new wine waste management method to create zero waste and recover all wine industry energy and resources (Psarommatis and Bravos 2022). One of the most prevalent wine wastes, grape seed, was treated using aqueous ethanol. After that, grape seed oil and polyphenols were recycled, and the solid leftovers were anaerobically digested to produce methane. The grape seed oil was found to have 82% unsaturated fatty acids and meet all vegetable oil standards, making it a high-value, safe food oil (Chen et al. 2022). Kumar Surendra and Bhatt Vardhan Harsh proposed zero waste. Few Indian and international towns and nations have achieved zero waste by establishing a system that focuses on garbage creation rather than waste addressal. Zero waste has succeeded with decentralized management. They wanted to find the key zero-waste indicators to evaluate zero-waste management systems. Waste specialists have discovered important markers for zero-waste management systems (Kumar and Bhati 2022). After a thorough literature analysis, zero-waste indicators were categorized into geo-administrative, sociocultural, managerial, economic, environmental, organizational, and policy domains. A total of 238 preliminary zero-waste indicators were distributed to 650 highly experienced trash experts worldwide for input. Research goals are as follows:

- To carry out a case study of top ten ZWM companies
- Relationship between ZWM and Sustainable Development Goals

5.3 TOP TEN ZWM COMPANIES

5.3.1 Subaru

Subaru recycles everything. Over 12 years, Subaru's U.S. and two Japanese operations haven't exported rubbish to landfills. Subaru's zero-waste policy benefits the environment and community but hurts profit. Subaru saved $1–2 million annually by becoming zero waste in the United States. About 96% of Subaru auto parts are reusable or recyclable. Subaru has a goal of being a trash-free corporation, and they want to help the National Park Service and other organizations across the globe accomplish the same by teaching them their successful zero-waste process (Ando et al. 1997).

5.3.2 Toyota

In 2013, Toyota became an inaugural U.S. Zero-Waste Building Council member. By 2015, Toyota North America reduced non-regulated trash by 96% via prevention, reuse, and recycling. The 96% garbage reduction saved over 900 million pounds from landfills. The U.S. Zero-Waste Building Council has certified 27 Toyota operations in North America as zero-waste locations. EPA (Environmental Protection Agency), recognized Toyota Motor as North America's Very Large Business, Waste Wise Partner of the Year for 2015 (Kedves 2023).

5.3.3 General Motors

GM said in 2016 that it operated 152 "zero-waste" plants. Employee hard work, recycling, repurposing, waste-to-energy conversion, and new product development were all mentioned as reasons for their success. According to reports, GM made $1 billion by recycling 2 million metric tons of garbage, like Subaru. GM reinvests its $1 billion savings into research and development of more efficient cars and technologies. Some of the air filters and engine cover insulation at GM are made from recycled water bottles from Flint, Michigan. General Motors has made Chevrolet Bolt battery covers into nest boxes for local animals. Those are greatly appreciated by the ducks (Suddaby et al. 2023).

5.3.4 Google

It seems to reason that such a forward-thinking corporation would embrace the zero-waste movement. As a first step toward its goal of zero waste as a corporation, Google is concentrating on its data centers. Last year, 6 of their 14 data centers sent no trash to landfills. Outside of its data centers, Google recycles or repurposes 86% of its total worldwide trash. The moonshot initiative at Google is an all-out attempt to put to good use unused data servers. The staff at Google is known for their creative reuse and recycling of workplace furnishings. Google, being as enormous as it is, has implemented many composting systems in an effort to lessen the amount of organic waste it generates (Ahmed et al. 2023).

5.3.5 Microsoft

More than 44,000 people work at Microsoft over 500 acres in Redmond, Washington's, 125 buildings. Microsoft's efforts to reduce waste have both environmental and economic implications due to the company's scale. Microsoft has created systems to divert 90% of its garbage from landfills, which contributes to its goal of reducing its carbon footprint.

Microsoft had the first U.S. Green Building Council-certified zero-waste building. The IT division at Microsoft created a power management system for all 160,000 PCs, reducing power utilization by 27%. Microsoft encourages remote work and provides mobility incentives to help employees reduce their carbon footprints (Malyuga 2023).

5.3.6 Sierra Nevada

Sierra Nevada shows that zero-waste beer is difficult. Due to its efforts, the U.S. Zero-Waste Council awarded its Chico, California, brewery platinum-level zero waste in 2014. Recognition goes beyond certification. Sierra Nevada saved $5 million by avoiding landfills with 99.8% to 100% of their waste. Livestock and dairy producers get 150,000 pounds of malted barley and 4,000 pounds of hops from Sierra Nevada. Their Chico, California, brewery has composted over 5,000 tons of organic waste utilizing the first U.S. Hot Rot composting system since 2010. Sierra Nevada uses Hot Rot compost in their crops and gardens and a Mills River, North Carolina, composting company (Weatherly and Lyons 2023).

5.3.7 New Belgium Brewing

The brewery that invented Fat Tyre Belgian Style Ale also pioneered zero-waste brewing, removing 99.9% of its waste from landfills. Both New Belgium Brewing and Sierra Nevada have achieved platinum-level zero-waste accreditation from the U.S. Zero-Waste Business Council. New Belgium searched its Fort Collins, Colorado, facility's more than 500 garbage receptacles for recycling and composting opportunities. Workers suggested innovative applications for all waste after the audit. New Belgium began composting organic waste from its water treatment plant in 2016, decreasing landfill waste (Castillo et al. 2023).

5.3.8 Fetzer Vineyards

Fetzer Vineyards, one of Mendocino County's largest wineries, has managed to stay sustainable since its founding. Fetzer also became the first vineyard to openly declare its greenhouse gas emissions via the climate register and set a goal to become net carbon positive by 2030. The U.S. Zero-Waste Council also recognized Fetzer Brewing Company as platinum-level zero waste. Fetzer Vineyards zero waste: The world's largest wine company, Fetzer, is a B Corporation. The first 100% renewable winery was California's Fetzer Winery. Instead of disposing of their effluent, Fetzer developed a means to clean it up so that it could be reused. This was accomplished with the assistance of billions of worms and bacteria that like feeding on waste products (Castillo et al. 2023).

5.3.9 Unilever

The efforts by Unilever to become a zero-waste corporation show that they are seizing the initiative. In point of fact, all of the company's non-hazardous trash is diverted from landfills at at least 240 of its facilities and 400 of its locations. These shifts had a significant effect on the company's finances. According to Unilever, their zero-waste initiative not only saved them over $225 million but also resulted in the creation of new employment. In an attempt to disseminate their approach of producing zero waste to as many companies as possible, Unilever teamed with 2degrees. The consumer goods company Unilever has committed to making all of its packaging recyclable, reused, or biodegradable by the year 2025. In order to solve the special problems associated with recyclable packaging, Unilever developed a brand new technology called the CreaSolv® Process (Akmal and Affandi n.d.).

5.3.10 Procter & Gamble

A significant zero-waste objective has been established by Procter & Gamble. They have set a goal for the corporation to transmit no trash from production to landfills by the year 2020. If this waste from production could be eliminated, Procter & Gamble would be able to cut down on the amount of trash they create by 95%. Their present level of development is significant, as 55% of their facilities no longer deposit any trash from production in landfills. In Hungary, the Always team at Procter & Gamble transfers manufacturing leftovers to a local cement firm, where they are burned to build bricks using the energy released from the combustion process (Esmaeilian et al. 2018). Composting the garbage generated at one of Procter & Gamble's facilities in China allows the company's "nutritional soil" to be utilized in the country's public parks. In order to make wall partitions for residences and workplaces, waste products from manufacturing are shredded and compressed in India. These partitions may then be used (Harivand 2023).

5.4 REGULATORY INITIATIVES

In order to achieve zero waste to landfill status, it is necessary to establish the necessary laws, policies, and programs and strictly adhere to them. It is expected of each and every participant to stick to a comprehensive strategy that includes certain dates. The first thing that has to be done is to alter the behaviors of people via civic education, with the primary focus being on altering their routines. People must embrace sustainable consumption. Unmanaged wastes harm climate change and human health. Many societies are hesitant to adopt sustainable lifestyles that would reduce environmental destruction. Sensitization, education, training, and new ZWM research must be funded by governments at all levels.

Non-biodegradable and hazardous product manufacturers must be held accountable for their environmental effects. To promote product take-back at end-of-life, producers must design effective take-back initiatives. Manufacturers must innovate to repair, rebuild, revive, reprocess, and remake returned items. Monetary incentives and discounts on new items are needed. This will encourage producers and consumers to take responsibility, decrease consumer dumping, and guarantee manufacturers make recyclable items (Yang et al. 2023).

Waste recycling and material recovery are poorly regulated in most nations. Some countries politicize waste management enforcement. For environmental sustainability, regulations and laws should target 100% recycling, 100% resource recovery, and zero landfills and incineration. Innovative technologies and methods are best for zero waste's cost-saving, economic, hygienic, environmental, waste management, material recycling, and resource recovery. Innovative eco-friendly technologies save money, reduce infections and contaminations, and allow metering and recordkeeping. Zero-waste methods may be time-consuming, deceptive, and complex. Waste collection, sorting, and conversion might cost more than product development. Converting and reusing garbage may also violate social, religious, and cultural norms (Farré et al. 2023).

5.5 ZWM AND SUSTAINABLE SOLUTION

India is the location of a number of manufacturing enterprises, many of which are long-established heritage names. In order to address the issue of climate change, they will need to adjust their perspective and devise a waste management strategy. There are many energy-wasting pieces of equipment that need to be changed. Then, low-hanging fruits in the supply chain may be easily replaced or eliminated (Ma et al. 2023).

CE and energy efficiency may help India reach net zero. A business-wide sustainable value chain is needed. Data and decarbonization regulations may improve results. Michel Fredeau, managing director and senior partner at Boston Consulting Group, France, recommended research and development (R&D) at the CII Manufacturing Conclave 2022. Changing a company model requires convincing consumers and skill development (Tripathy and Dastrala 2023). The business model might incorporate recycling units for net-zero items. This may be a chance. India wastes 62 million tons every year. Reusing garbage requires creativity. The business might be self-sustaining. "These broad-based thoughts need to incorporate solutions to make allied elements beneficial," said JBM Group vice chairman and Linde Wiemann Gmbh chairman Nishant Arya (Debele et al. 2023).

Green hydrogen projects can collect carbon for clean production. Fix green travel obstructions. The distribution system's fragments must be reunited. Energy policies require structure. R Mukundan, Tata Chemicals Limited CEO and MD, said natural gas might replace petrol. Solar, wind, and hydrogen energy should be mainstream for the environment. Technology expenditures may assist in meeting industrial objectives like greenhouse gas reduction. Include sustainability to boost manufacturing (Singh and Amist 2023).

Manufacturing, government, and nongovernmental organizations will share environmental responsibilities. Manufacturers may start green initiatives with stakeholder confidence. Green manufacturing and vapor-absorption equipment increased vehicle economy. We want to recruit top personnel and achieve product recyclability and a circular supply chain. Apurva Gupta, Hero MotoCorp's chief sustainability officer, said, "We enroll dealerships, suppliers, and customers." Sewage water and zero liquid discharge plants are green efforts. The firm developed greenhouses with green roofs. Hydroponics reduces water use to 2% of agriculture. The greenhouse recycles carbon dioxide, boosting photosynthesis. Green walls oxygenate the workplace (Suarez Sánchez and Jaimes 2023).

As industries go green, waste collection technology should be used. R&D specialists who analyze sustainable manufacturing technologies have several business options here. Tech and data may reduce supply chain waste. Carbon neutrality will cost more, but technology can adapt to new energy. VP Gauri A. Kirloskar stated, "Kirloskar Oil Engines Ltd has installed a plant unit that can convert 100kg of plastic waste to fuel plant, whose characteristics are similar to diesel" (Singh et al. 2023).

Manufacturing Industry 4.0 requires sustainability-focused capital investment. These investments may not pay off immediately, so they may be small. Search for low-interest capital-friendly policies. Data will inform Manufacturing 4.0 processes and customer preferences. Cybersecurity must support this. Climate change requires technology and infrastructural preparation (Gani and Fernando 2023). Market incentives for carbon investments might help solve climate change. Ashwath Ram, Cummins India Ltd.'s managing director, said, "Planet 2050 is our carbon neutral goal" (Maurya et al. 2023). Employee awareness and resource usage are promoted by PLANET 2050. To decrease waste, reuse resources, and neutralize water, this agenda redesigns processes and goods. Community collaboration and greenhouse gas reduction are included.

Technological advancements enhance business. Profitability and scalability grow with sustainability. Technology sustains industry. Low-carbon manufacturing attracts worldwide customers. It must be inexpensive however. Whenever feasible, use green tech. CEO and MD of Dalmia Cement (Bharat) Ltd. and COP26 business leader Mahendra Singhi urged green finance and strategies to make India a green manufacturing hub (Maurya et al. 2023).

Countries might enhance the global economy at COP26. Countries may approach COP26 differently. Kirloskar stated, "Globally, no nation should win." Trash management relies on trash recovery to minimize waste volume and avoid landfill overflow. Informal recycling helps control garbage and employs low-income individuals. Waste management efficiency promoted community engagement in several places. Recent public awareness and engagement reduced waste at the source and made people more ecologically sensitive (Shah et al. 2023). The zero-waste hierarchy outlines a series of policies and techniques that may be implemented to facilitate the zero-waste system from the most efficient and effective to the least resource-intensive. It's designed to be useful for everyone from government officials to business owners to regular folks. This document aims to add nuance to the 5Rs (Refuse, Reduce, Reuse, Repurpose, Recycle) as shown in

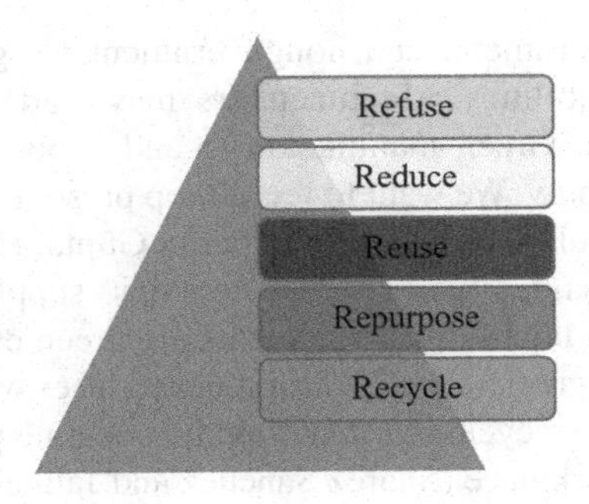

FIGURE 5.3 Five R's of zero-waste hierarchy.

Figure 5.3, promote policy, activity, and investment at the top, and provide a road-map for those who want to create zero-waste systems or products. It helps zero-waste planners by providing a road map and assessment framework for solutions. Users are urged to work from the top down when formulating strategies and plans (Snoussi et al. 2023).

The United Nations (UN) has adopted 17 Sustainable Development Goals (SDGs), often known as the Global Goals for Sustainable Development. The UN also aims to accomplish these 17 important goals before it's too late to avert the climate calamity (Yao et al. 2023). In addition, to guarantee our children and grandchildren a brighter, more environmentally friendly, and more pristine future, every single one of the 17 SDGs is important and vital. However, the importance of recycling underscores the significance of the 12th goal, which is to ensure responsible consumption and production (Sharma et al. 2023).

As stated by the UN, we all need to work together to make sure that these objectives are met by the year 2030. Because of this, the UN strongly encourages people, governments, and organizations to recycle, cut waste, and work toward the goal of producing zero waste. Because the 12th SDG emphasizes the significance of responsible consumption and production, it also emphasizes the value of recycling (Ding et al. 2023).

The restructuring of industrial production is one of the major challenges that must be addressed on the path to sustainable development. Therefore, many expectations are pinned on the part that Industry 4.0 will play (Chiarini et al. 2020). Research linking factory digitalization to the UN's 2030 Agenda for Sustainable Development is scarce. Industry 4.0 should prioritize environmental issues for sustainable growth. This means that essential components of environmentally friendly production must be included in the very concept of digital manufacturing (Yang and Yan 2023).

The importance of recycling is continuously rising in the minds of people everywhere. It is crucial at this time to stress the relationship between recycling and being environmentally conscious. To begin, you might say that recycling is a sustainable activity. This is why the value of recycling is being more recognized. Recycling and environmental protection are two subjects that go hand in hand.

SDGs can't be achieved without recycling. This guarantees long-term viability. Garbage, for instance, is a major cause of the present climate crisis; hence, all institutions, including governments, need efficient and adaptable waste management systems. In addition to these additional options for zero-waste management, recycling is an efficient waste management approach (Cole et al. 2014).

5.6 INDUSTRY REVENUE IN INDIA FROM 2012 TO 2024

Prosperity generally correlates with rubbish production. By 2050, daily per capita waste generation is expected to climb by 19% in high-income countries and 40%–50% in low- and middle-income countries. India's waste collection, treatment, disposal, and materials recovery income is expected to reach $3.5 billion by 2024.

5.7 ZWM IMPACT ON THE INDIAN ECONOMY

ZWM uses value chain-wide waste elimination technologies and systems to support circular economies. ZWM protects the environment via an inclusive, ecological, and equitable shift. Preventing biodiversity loss and soil quality deterioration requires quick reactions to unexpected temperature changes and season uncertainty. Zero-waste recycles till optimal consumption is reached [56, 57].

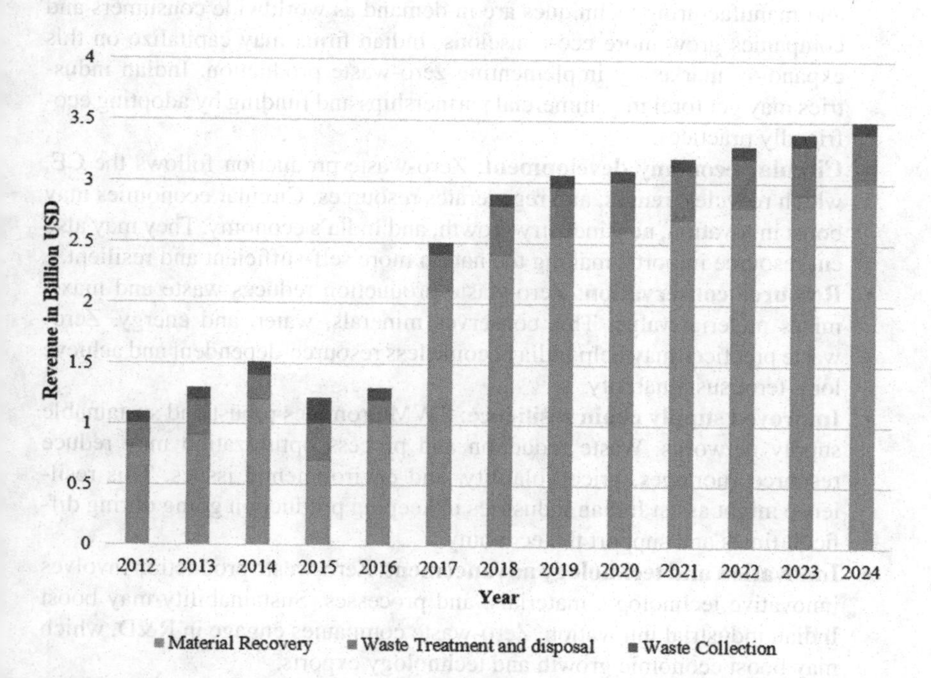

FIGURE 5.4 Revenue in billion USD of the Indian industry (2012–2024).

Zero-waste production is vital for environmental, economic, and personal reasons. The impact of ZWM on the Indian economy is presented below:

- **Cost savings**: ZWM prioritizes resource efficiency and minimizes waste. Companies may reduce procurement expenses by reducing raw material usage and optimizing manufacturing processes. This boosts profitability, cost competitiveness, and return on investment.
- **Increased productivity**: Zero-waste production frequently involves process optimization and lean manufacturing. Reducing downtime, faults, and operations may boost productivity. Productivity increases production, efficiency, and economic output.
- **Environmental benefits**: Zero-waste production emphasizes recycling and reuse. This method reduces pollutants, greenhouse gas emissions, and resource utilization. India's population and development pose environmental issues. Zero-waste practices may improve human health and natural ecosystems by cleaning up air, water, and land.
- **Job creation and skill development**: Zero-waste production demands new technologies, techniques, and capabilities. This transition generates career and skill development possibilities across industries. Sustainable manufacturing training can empower India's large workforce and boost their green economy employment.
- **Market demand and international competitiveness**: Sustainable goods and manufacturing techniques are in demand as worldwide consumers and companies grow more eco-conscious. Indian firms may capitalize on this expanding market by implementing zero-waste production. Indian industries may get foreign commercial partnerships and funding by adopting eco-friendly practices.
- **Circular economy development**: Zero-waste production follows the CE, which recycles, reuses, and regenerates resources. Circular economies may boost innovation, new industry growth, and India's economy. They may also cut resource imports, making the nation more self-sufficient and resilient.
- **Resource conservation**: Zero-waste production reduces waste and maximizes material value. This conserves minerals, water, and energy. Zero-waste practices may help India become less resource-dependent and achieve long-term sustainability.
- **Improved supply chain resilience**: ZWM promotes robust and sustainable supply networks. Waste reduction and process optimization may reduce resource shortages, price volatility, and environmental issues. This resilience might assist Indian industries in keeping production going during difficult times and support the economy.
- **Innovation and technology advancement**: Zero-waste production involves innovative technology, materials, and processes. Sustainability may boost Indian industrial innovation. Zero-waste companies engage in R&D, which may boost economic growth and technology exports.

- **Enhanced corporate social responsibility (CSR)**: ZWM promotes sustainability and environmental care. Indian firms may attract socially concerned customers and investors by adopting zero-waste practices. This good image may boost market share, brand value, and company longevity.
- **Reduced waste management costs**: India has problems with landfill space, waste treatment facilities, and pollution. Zero-waste production reduces trash and waste management expenses for enterprises and the government. This frees up funds for infrastructure, health care, and education.
- **Export potential**: ZWM may make Indian firms sustainable leaders. This might increase exports, particularly to nations with strict environmental restrictions and a rising demand for eco-friendly items. Zero-waste Indian enterprises may capitalize on the worldwide demand for ecological and socially responsible goods.

ZWM may boost economic development and sustainability in India. However, the government, industry stakeholders, and the public must work together to reduce hurdles, raise awareness, and offer assistance and infrastructure to achieve universal zero-waste adoption (Khalid et al. 2023).

5.8 CONCLUSION

As the world's economy and population have expanded at a dizzying rate, so too has the mountain of garbage that has accumulated. The potential for adverse consequences on the environment and the consequent waste of resources has captivated the attention of people everywhere. Illegal dumping and transboundary movement of industrial waste, informal recycling of electronic trash, food waste, greenhouse gas emissions, and excessive packaging are all instances of these impacts. The waste management issues we face now are far more serious than they were in the past, and finding a workable solution is a major challenge for everyone on Earth. When applied to the problems associated with solid waste management, the "zero-waste" concept becomes a potent instrument. The objective of "zero waste" is to have people thinking differently about resource cycles so that everything can be recycled or repurposed. To help the reader better grasp the notion of zero waste, this study outlined the most prevalent zero-waste methods. There are now numerous large-scale programs being implemented in cities, businesses, and individual homes, all of which provide many useful suggestions for us to eventually achieve zero waste.

REFERENCES

Aqilah, N.M.N., et al., A review on the potential bioactive components in fruits and vegetable wastes as value-added products in the food industry. *Molecules*, 2023. 28(6): p. 2631.

Ahmed, F., et al., A conceptual framework for zero waste management in Bangladesh. *International Journal of Environmental Science and Technology*, 2023. 20(2): p. 1887–1904.

Akmal, D.M. and R.A. Affandi, Integrating Government Policy and MNC Actions in SDGs: a Case of Jakarta Government and Unilever Indonesia, *5th European International Conference on Industrial Engineering and Operations Management*, Rome, Italy, July 26–28, 2022.

Ali, S. and F. Shirazi, The Paradigm of Circular Economy and an Effective Electronic Waste Management. *Sustainability*, 2023. 15(3): p. 1998.

Ando, A., et al. VUV and soft X-ray light source" New SUBARU". in *Proceedings of the 1997 Particle Accelerator Conference (Cat. No. 97CH36167)*. 1997. IEEE.

Boadway, R. and F. Flatters, The taxation of natural resources: principles and policy issues, in *Taxing Choices for Managing Natural Resources, the Environment, and Global Climate Change: Fiscal Systems Reform Perspectives*. 2023, Springer. p. 17–81.

Burke, G.T., et al., Making Space for Garbage Cans: How emergent groups organize social media spaces to orchestrate widescale helping in a crisis. *Organization Studies*, 2023. 44(4): p. 569–592.

Castillo, A.B., et al., Bioethanol production from waste and nonsalable date palm (*Phoenix dactylifera* L.) fruits: Potentials and challenges. *Sustainability*, 2023. 15(4): p. 2937.

Chen, J., et al., New insights into wine waste management: Zero waste discharge-driven full energy/resource recovery strategy. *Results in Engineering*, 2022. 15: p. 100606.

Chiarini, A., V. Belvedere, and A. Grando, Industry 4.0 strategies and technological developments. An exploratory research from Italian manufacturing companies. *Production Planning & Control*, 2020. 31(16): p. 1385–1398.

Cole, C., Osmani, M., Quddus, M., Wheatley, A., & Kay, K. Towards a zero waste strategy for an English local authority. *Resources, Conservation and Recycling*, 2014. 89: p. 64–75.

Debele, G.J., F.U. Fita, and M. Tibebu, Prevalence of ABO and Rh blood group among volunteer blood donors at the blood and tissue bank service in Addis Ababa, Ethiopia. *Journal of Blood Medicine*, 2023: p. 19–24.

Debnath, B., et al., Modelling the barriers to sustainable waste management in the plastic-manufacturing industry: An emerging economy perspective. *Sustainability Analytics and Modeling*, 2023. 3: p. 100017.

Ding, B., X. Ferras Hernandez, and N. Agell Jane, Combining lean and agile manufacturing competitive advantages through Industry 4.0 technologies: an integrative approach. *Production Planning & Control*, 2023. 34(5): p. 442–458.

Farooqi, Z. U. R., Kareem, A., Rafi, F., & Ali, S. (2021). Solid Waste, Treatment Technologies, and Environmental Sustainability: Solid Wastes and Their Sustainable Management Practices, in *Handbook of Research on Waste Diversion and Minimization Technologies for the Industrial Sector* (pp. 35–57). IGI Global.

Farré, J.A., et al., Pneumatic Urban Waste Collection Systems: A Review. *Applied Sciences*, 2023. 13(2): p. 877.

Fragapane, G., et al., A global survey on the current state of practice in Zero Defect Manufacturing and its impact on production performance. *Computers in Industry*, 2023. 148: p. 103879.

Gale, J., Reduce, reuse, recycle: In that order. *Clinical & Experimental Ophthalmology*, 2023.

Gani, A.B.D. and Y. Fernando, Digital empathy and supply chain cybersecurity challenges: concept, framework and solutions for small-medium enterprises. *International Journal of Management Concepts and Philosophy*, 2023. 16(1): p. 1–10.

Esmaeilian, B., Wang, B., Lewis, K., Duarte, F., Ratti, C., & Behdad, S. (2018). The future of waste management in smart and sustainable cities: A review and concept paper. *Waste Management*, *81*, 177–195.

Guandalini, I., Sustainability through digital transformation: A systematic literature review for research guidance. *Journal of Business Research*, 2022. 148: p. 456–471.

Harivand, R.G., The Future of IO Kids: Establishing Brushing Habit in Early Childhood. 2023, Purdue University Graduate School.

Kedves, P.E., Toyota—Looming Downfall in an Electrifying Industry, in *Overcoming Crisis: Case Studies of Asian Multinational Corporations*. 2023. Parissa Haghirian, Editors, p. 171–183.

Kerdlap, P., J.S.C. Low, and S. Ramakrishna, Zero waste manufacturing: A framework and review of technology, research, and implementation barriers for enabling a circular economy transition in Singapore. *Resources, Conservation and Recycling*, 2019. 151: p. 104438.

Khalid, M. Y., Arif, Z. U., Hossain, M., & Umer, R. Recycling of wind turbine blades through modern recycling technologies: A road to zero waste. *Renewable Energy Focus*, 2023. 44: p. 373–389.

Kumar, S. and H.V. Bhati, Chapter 4 – Waste management to zero waste: Global perspectives and review of Indian law and policy, in *Emerging Trends to Approaching Zero Waste*, C.M. Hussain, S. Singh, and L. Goswami, Editors. 2022, Elsevier. p. 79–101.

Kumari, T. and A.S. Raghubanshi, Waste management practices in the developing nations: challenges and opportunities. *Waste Management and Resource Recycling in the Developing World*, 2023: p. 773–797.

Lušić, M., et al., Towards Zero Waste in Additive Manufacturing: A Case Study Investigating one Pressurised Rapid Tooling Mould to Ensure Resource Efficiency. *Procedia CIRP*, 2015. 37: p. 54–58.

Ma, W., et al., Evaluating carbon emissions of China's waste management strategies for building refurbishment projects: Contributing to a circular economy. *Environmental Science and Pollution Research*, 2023. 30(4): p. 8657–8671.

Makinde, O., T. Munyai, and E. Nesamvuni. A Hybrid Structural Interaction Matrix Approach to Prioritise Process Wastes Generated in a Manufacturing Organisation. in *Manufacturing Driving Circular Economy: Proceedings of the 18th Global Conference on Sustainable Manufacturing*, October 5–7, 2022, Berlin. 2023. Springer.

Malyuga, E.N., A corpus-based approach to corporate communication research. *Russian Journal of Linguistics*, 2023. 27(1): p. 152–172.

Maurya, S., et al., Conceptual Design of Extrusion Systems for Cement Paste 3D Printing, in *Low Cost Manufacturing Technologies: Proceedings of NERC 2022*. 2023, Springer. p. 59–71.

Meng, F., et al., Planet-compatible pathways for transitioning the chemical industry. *Proceedings of the National Academy of Sciences*, 2023. 120(8): p. e2218294120.

Naveenkumar, R., et al., A strategic review on sustainable approaches in municipal solid waste management and energy recovery: Role of artificial intelligence, economic stability and life cycle assessment. *Bioresource Technology*, 2023: p. 129044.

Nguea, S., Improving human development through urbanization, demographic dividend and biomass energy consumption. *Sustainable Development*, (2023), *31*(4), 2517–2535.

Pérez, F., et al. A CPPS Architecture approach for Industry 4.0. in *2015 IEEE 20th conference on emerging technologies & factory automation (ETFA)*. 2015. IEEE.

Powell, D., et al., Advancing zero defect manufacturing: A state-of-the-art perspective and future research directions. *Computers in Industry*, 2022. 136: p. 103596.

Pozzi, R., T. Rossi, and R. Secchi, Industry 4.0 technologies: Critical success factors for implementation and improvements in manufacturing companies. *Production Planning & Control*, 2023. 34(2): p. 139–158.

Psarommatis, F. and G. Bravos, A holistic approach for achieving Sustainable manufacturing using Zero Defect Manufacturing: a conceptual Framework. *Procedia CIRP*, 2022. 107: p. 107–112.

Psarommatis, F., et al., Zero defect manufacturing: state-of-the-art review, shortcomings and future directions in research. *International Journal of Production Research*, 2020. 58(1): p. 1–17.

Sengupta, S. and H. Dreyer, Realizing zero-waste value chains through digital twin-driven S&OP: A case of grocery retail. *Computers in Industry*, 2023. 148: p. 103890.

Shah, S.A.R., et al., Technology, urbanization and natural gas supply matter for carbon neutrality: A new evidence of environmental sustainability under the prism of COP26. *Resources Policy*, 2023. 82: p. 103465.

Sharma, M., et al., Green logistics driven circular practices adoption in industry 4.0 Era: A moderating effect of institution pressure and supply chain flexibility. *Journal of Cleaner Production*, 2023. 383: p. 135284.

Singh, A., K. Singh, and J. Singh, An Evaluation of the Performance of Different Biodiesel Fuels on Engine Efficiency and Emission Characteristics, in *Emerging Trends in Mechanical and Industrial Engineering: Select Proceedings of ICETMIE 2022*. 2023, Springer. p. 297–308.

Singh, K.K. and A.D. Amist. Measuring Effectiveness of CSR Activities to Reinforce Brand Equity by Using Graph-Based Analytics. in *Advances in Cognitive Science and Communications: Selected Articles from the 5th International Conference on Communications and Cyber-Physical Engineering (ICCCE 2022)*, Hyderabad, India. 2023. Springer.

Snoussi, Y., et al., Green, zero-waste pathway to fabricate supported nanocatalysts and anti-kinetoplastid agents from sugarcane bagasse. *Waste Management*, 2023. 155: p. 179–191.

Sohail, M., et al., Agricultural communities' risk assessment and the effects of climate change: A pathway toward green productivity and sustainable development. *Frontiers in Environmental Science*, 2023. 16648714: p. 123.

Suarez Sánchez, M.Á. and J.D. Diaz Jaimes, Plan De Marketing Aplicado A La Empresa Hero Motocorp Para El Posicionamiento De Marca En El Departamento De Santander. 2023.

Suddaby, R., L. Manelli, and Z. Fan, Corporate purpose: A social judgement perspective. *Strategy Science*, 2023. 8(2), 202–2011.

Tan, Q., et al., Is reusable packaging an environmentally friendly alternative to the single-use plastic bag? A case study of express delivery packaging in China. *Resources, Conservation and Recycling*, 2023. 190: p. 106863.

Tripathy, A. and S.M. Dastrala, Make in India: So Far and Going Ahead. IIM Bangalore Research Paper, 2023(674).

Turok, I. and G. McGranahan, Urbanization and economic growth: the arguments and evidence for Africa and Asia. *Environment and Urbanization*, 2013. 25(2): p. 465–482.

Vinti, G., et al., Health risks of solid waste management practices in rural Ghana: A semi-quantitative approach toward a solid waste safety plan. *Environmental Research*, 2023. 216: p. 114728.

Weatherly, S. and R. Lyons, The photolytic conversion of 4-nonylphenol to 4-nonylcatechol within snowpack of the Palisade Glacier, Sierra Nevada, CA, USA. *Science of the Total Environment*, 2023. 876: p. 162835.

Weber, S., et al., Textile waste in Ontario, Canada: Opportunities for reuse and recycling. *Resources, Conservation and Recycling*, 2023. 190: p. 106835.

Wikurendra, E.A., N.S. Abdeljawad, and I. Nagy, A review of municipal waste management with zero waste concept: Strategies, potential and challenge in Indonesia. *International Journal of Environmental Science and Development*, 2023. 14(2): 147–154.

Yang, C. and X. Yan, Impact of carbon tariffs on price competitiveness in the era of global value chain. *Applied Energy*, 2023. 336: p. 120805.

Yang, G., et al., How does the "Zero-waste City" strategy contribute to carbon footprint reduction in China? *Waste Management*, 2023. 156: p. 227–235.

Yao, Z., W. Qi, and J.L. Francisco Alves, Editorial for the Special Issue on the Environmentally Friendly Management and Treatment of Solid Waste to Approach Zero Waste City. 2023, MDPI. p. 826.

6 Frugality and Sustainability

Essential Concepts for Sustainable Manufacturing

Rasleen Kour
Indian Institute of Technology Ropar, Rupnagar, India

6.1 INTRODUCTION

The Peach Blossom Spring, one of the well-known works by a prominent Chinese poet, narrates the tale of a fisherman who accidentally reaches a secret society outside of the world and gets lost. The remote location is a joyful, lovely, serene area covered in fragrant peach blossoms. After making use of their hospitality for a while, the fisherman leaves to reveal the whereabouts of the stunning location to everyone. He forgets the route on the way home and never makes it back. Sarokin (2022) spoke about how the outer world has once more forgotten this Peach Blossom Spring. Even when humanity first manifested, it was in the idyllic Garden of Eden, the Garden of Paradise. The original Garden of Paradise could withstand any catastrophe. Consider the 2004 tsunami in India (on the islands of Andaman and Nicobar), which demonstrated the power of nature to deal with any calamities (Ghosh, 2020). The study shows that the Rhizophora mangroves were undamaged by the abrupt disruption. The area where Rhizophora mangroves were present as seaward-facing trees demonstrated resilience to the sudden upsurge in sea level. The 2004 tsunami severely damaged the places where human activity had already deteriorated the ecosystem. The incident highlights the value of mangrove protection and restoration since these ecosystems can significantly lessen the effects of climate change and natural catastrophes on coastal regions (Sarokin, 2022). As technology advances, restoring the lost paradise has become a utopian dream. Nowadays, there is a growing urge to conserve the paradise that has been lost. This chapter discusses how this need can be met through sustainability. The goal of sustainability is to meet the demands of the present generation without jeopardizing those of the future (Sarokin, 2022). Sustainability emphasizes using less energy and resources, producing less waste, employing fewer hazardous materials, and manufacturing more environmentally friendly goods. As a result, conserving resources and using them wisely/frugally are essential to sustainable growth.

DOI: 10.1201/9781003309123-6

This chapter will review how being frugal is a crucial component of sustainability. Additionally, it will shed light on the beneficial relationship between sustainability and frugality and the role that frugal practices play in achieving sustainable goals.

6.2 SUSTAINABILITY

The Latin root of sustainability is *sustinere*, which means to uphold, support, or keep (Online Etymology Dictionary [Sustainable, 2022]). It means the capacity for sustained endurance. Earlier, sustainability was only defined in terms of environmental sustainability. It was used to preserve natural resources to be reserved for future generations. Hans Carl Von Carlowitz coined the term "sustainability" (German: Nachhaltigkeit) in 1713, intending to safeguard forests (Wikipedia [Sustainability, 2023]). Soon, it included the protection of food, animals, and plants. In technical terms, sustainability was used in the 1972 top magazine *Blueprint for Survival* (Sarokin, 2022). As a result, the definition of sustainability expands to include lifestyle choices, population management, and the conservation of natural resources. The Brundtland Commission of the United Nations (UN) gave a detailed description of the term "sustainability" in 1987, stating that it meant "to meet the needs of the present [generation] without comprising the ability of future generations to meet their own needs" (Sarokin, 2022; Imperatives, 1987). At the 2005 World Summit on Social Development, the definition of sustainability changed once again. It consists of three components: people (social commitments), the planet (environmental conservation), and profits (corporate profits) (Sarokin, 2022). Therefore, the three-legged stool represents the sustainability goals: environmental, societal, and economic. It deals with all the issues related to economic equity, finance, social justice, and human dignity. The goal was to incorporate the problems of economic equity, finance, social justice, and human dignity. Talking about sustainability is crucial because the impact of technological developments on nature and climate has grown and cannot be reversed.

6.2.1 Why Is There a Need to Talk About Sustainability?

6.2.1.1 Climate Change

With the 19th century beginning, scientists started discussing the effects of modernity on the environment. Svante Arrhenius (1896), a Swedish chemist, hypothesizes that CO_2 could have a warming influence on the planet's ability to regulate its temperature. By the 1950s, there were growing fears as popular newspaper stories warned that emissions of fossil fuels would disrupt the environment's carbon dioxide balance and lead to widespread global warming and drastic climate change. In addition, there were other frightening reactions, such as the peak oil theory, which questions the limits of petroleum that can be extracted, and Rachel Carson's *Silent Spring*, which discusses the dangers of pesticides (Sarokin, 2022). Additionally, the book *The Population Bomb* forewarned of the catastrophic impact that an increasing human population will have on the planet's resources (Ehrlich and Ehrlich, 2009). The Limits of Growth Report from 1972, which uses computer modeling to illustrate global forecasts, warns that everything will run out within the next 100 years (Meadows et al., 1972). Schumacher (1973), who believes that tiny is beautiful and refers to society's unsustainable path as "conducting the economic affairs of man as if people really did not matter at all," brought attention to the issues of limited resources and excessive growth. The same year, the world went through its first "energy crisis" as petrol and other resources started to become less (Sarokin, 2022). The issue is reaching new heights as we approach the 20th century. Jared Diamond talks about his predictions for the future in his book *Upheaval* in a 2019 interview:

> I would estimate the chances are about 49 percent that the world as we know it will collapse by about 2050.... At the rate we're going now, resources that are essential for complex societies are being managed unsustainably...by 2050 either we've figured out a sustainable course, or it'll be too late.
>
> *(Sarokin, 2022; Wallace-Wells, 2019)*

This demonstrates that sustainability is the only way to manage the earth's resources and a good existence.

Moreover, the Tasmanian tiger and the passenger pigeon are extinct, and we must accept that the loss is irreparable (Sarokin, 2022). Natural resources on earth are finite and will run out if not used sustainably. "We have not inherited the earth from our ancestors; rather, we have borrowed it from our children," as the International Union for Conservation of Nature, a Swiss organization, famously said. The most significant Brundtland Commission final report, titled "Our Common Future," lists every global disaster that resulted in thousands of fatalities and made many more homeless, including the drought crisis in Africa, the Bhopal gas tragedy in India, the liquid gas tanks tragedy in Mexico City, the Chernobyl nuclear reactor's aftereffects in Europe, the flow of agricultural chemicals, and mercury into the Rhine River (Imperatives, 1987). The Brundtland Commission report also discusses the role of technology and social organization in regulating economic growth. It also addresses issues such as the basic needs of people with low incomes and improvement of their quality of life. The first Earth Summit took place in 1992, and 172 nations participated to demonstrate their support for the cause of sustainability. The UN Sustainable Development Goals were released in 2015 after extensive international discussion, focusing on 17 target areas (no poverty, zero hunger, good health and well-being, quality education, gender equality, clean water and sanitation, affordable and clean energy, economic growth, innovation, infrastructure, reduced inequalities, responsible consumption and production, climate action, life below water, life on earth, peace, justice, and strong institutions, partnerships for the goals) (17 Goals, n.d.). Many countries, especially China and India (also pioneers of frugality), are significantly working on these goals to reduce global poverty and other socio-economic problems. Nearly one billion people have been elevated out of extreme poverty (Sarokin, 2022). It is a truly remarkable and inspiring accomplishment made possible through frugal innovations.

6.2.1.2 Dilemma of Social Control

Another reason to talk about sustainability and frugality is the effect of recent technology on society and the dilemma of social control. David Collingridge discusses the dilemma of social control regarding the challenge of regulating technological innovation (Genus and Stirling, 2018). The dilemma arises from the fact that at the early stages of technology development, its impacts on society may be uncertain or unpredictable. At the same time, later, it may be difficult or costly to control or alter its trajectory. According to Collingridge, there are two stages in technology development: The "pre-implementation" phase, when the technology is still in the laboratory or design stage and the "post-implementation" phase, when the technology has been widely adopted and integrated into society. In the pre-implementation phase, it is not easy to assess the social and environmental impacts of the technology, and hence, it may be challenging to regulate or control its development. In contrast, in the post-implementation phase, the effects of the technology may be more apparent, but it may be challenging to modify or regulate the technology since it has already become entrenched in society. Collingridge argues that this dilemma can be addressed by adopting "a responsive approach" to technology regulation, which involves monitoring the technology's impacts and adjusting regulating measures accordingly. It means we need to consider the effect of technology at an early stage (design stage) and design it using sustainable and frugal principles that help to avoid unnecessary manufacturing. Sustainable and frugal innovations can help create the technology in the most reasonable ways.

6.3 FRUGALITY

Frugality derives from the Latin word *frux*, which means fruit or value (Frugal, 2020). The synonyms of frugality are thrifty and economical. Being frugal involves managing resources and money with care and consideration. It entails making deliberate decisions to avoid waste, pointless spending, and excessive consumption. Frugal individuals prioritize saving and investing for the future rather than spending money on things that provide short-term pleasure or gratification. In many cultures, frugality is frequently regarded as a virtue essential to living a satisfied life (Antonio, 2010).

Frugality is a way of life that stresses the importance of thriftiness and has long been linked to social and environmental concerns. Nonetheless, the decline of frugality has become a pressing issue in contemporary society, where people tend to prioritize their spending and technological excess without thinking about others (including the environment). Frugality has historical roots in ancient traditions (Antonio, 2010). The philosophies such as Epicurus, Aristotle, and Aquinas advocated for self-discipline and mastery of the senses. Frugality was regarded as a virtue of spiritual discipline and a responsibility to promote sustainability, compassion, and benevolence toward others. However, the ascent of capitalism and technological advancements has led affluent societies to view frugality as an obstacle to prosperity, thus discouraging its practice (Nash, 1995). But now, less consumption is what we require for sustainability, and the frugal offers solutions to satisfy everyone's requirements without harming the environment. Frugality principles are very similar to sustainability, such as cost-effectiveness, improving marginal communities, mitigating resource scarcity, quality production, user-oriented, aesthetically enriching, and long-term achievement (Leliveld and Knorringa, 2018). Though sustainability and frugality have different ways of implementing their values, this chapter argues that if they collaborate, it will yield better outcomes for society.

6.4 THE CONNECTION BETWEEN SUSTAINABILITY AND FRUGALITY

There is a strong link between frugality and sustainability, as both complement each other. For example, a frugal lifestyle can help reduce waste and consumption, contributing to a more sustainable future. On the other hand, sustainability principles encourage frugal living, such as reducing energy usage or choosing environmentally friendly products to save costs. Frugal innovations also help fulfill three dimensions (ecological, social, and economic) of sustainability in developed and developing nations. Simoes et al. (2018) and Akbar and Subramaniam (2019) discuss various indicators that show how frugal innovation developers assist in the social sustainability performance of the product; stakeholders (including local community users for better outputs), social assessment areas (effect on society and product responsibility), frugal characteristics (functional, robust, user-friendly, growing, affordable and local), and social value/impact (how the product will influence or impact the well-being of stakeholders). Frugal innovation in developing countries has clear social implications (low price yet good quality products), economic impact (poverty reduction through local markets), and ecological impact (use of clean, renewal, and local resources). Frugal innovations also help improve sustainability performance in developed nations through reverse innovations (from developing to developed countries). Weyrauch and Herstatt (2016) and Albert (2019) define frugal as innovations that focus on three criteria: substantial cost reductions, concentration on core functionalities, and optimized performance level. So, when we define frugal innovations in terms of economical and responsible ways to provide good environmental and social impacts, it is always related to sustainability. Some believe that frugal innovations can fulfill sustainability aims (at social, economic, and ecological balance) and stifle capitalist exploitation and inequality in developing countries (Akbar and Subramaniam, 2019).

- **Frugal innovation and ecological sustainability**: Frugal innovation helps ecological sustainability by providing eco-friendly, green, clean products that address environmental constraints. Their focus is to balance the planet's climate, mitigate the crisis, preserve environmental integrity, enhance the green supply chain, lower adverse ecological effects, wisely use natural resources, and manage the outcomes of any product. The benefit of using frugal innovation is that it uses economic/limited/less/saved/minimized energy and resources. It always goes for local and renewable resources, is robust, has a no-frills design, creates value from waste (means using waste as resources), reuses the resources, and believes in repairing and recycling rather than going for new ones. It means frugal innovation has a low impact on the planet, a low carbon footprint, and fewer emissions. Thus, it is similar to sustainability, focusing on green, clean, recyclable, and eco-friendly products.

- **Frugal innovation and social sustainability**: Frugal innovation is believed to be one of the practical steps in realizing social sustainability by resolving various societal problems by providing affordable solutions. It motivates equity and social justice, empowers the bottom-of-the-pyramid (BOP) people, and ends world hunger, social integrity, and social inclusiveness. It opens jobs and employment for underserved populations (primarily locals and women). It creates cost-saving opportunities for the BOP and provides products and services in different sectors, such as food, health, communication, technology, water, and transportation, without compromising the quality. Therefore, positive relations can be seen in frugal innovation and sustainability regarding affordability for poor people, such as providing shelters, jobs, and quality of food and water.
- **Frugal innovation and economic sustainability**: Frugal innovation always proves to be the fastest growing market in developing nations and also provides an opportunity for western frugal innovators to generate profit and survive in the crisis. Frugal innovation focuses on higher products by reducing the cost of production and distribution. Therefore, there is a similarity between frugal innovation and economic sustainability in terms of providing higher customer values, increasing the company's competitiveness and sales revenues, and thus innovation inherently.

Various successful examples among different cultures of how frugality can help reduce waste and promote ecological, economic, and social sustainability are as follows.

6.4.1 Successful Example of Contribution of Frugal Innovations in Promoting Sustainability

6.4.1.1 Sparky Dryer

In Uganda, an East African nation, most food gets wasted after harvest, and almost 50% of the food spoils before it is sold in the market (British Council, 2018). Kampala, the capital of Uganda, is the most significant source of solid waste generation, with an exponential increase from 407,890 tons in 2011 to 785,214 tons in 2017. It is the primary source of solid organic and biodegradable garbage and landfills, making it the second largest contributor to greenhouse gas emissions. The major problem is not the lack of food but storage facilities. People produce more food than they consume, but it gets spoiled quickly because of the weather, lack of electricity, and other resources. According to the UN World Food Program, one in three Ugandan school-age children does not have enough food to consume. Lawrence Okettayot, a graduate student in engineering, creates a low-tech thermos dehydrator dubbed the Sparky Dryer as a very effective means

FIGURE 6.1 Sparky Dryer.

Source: Apio (2021). https://itsmapio.com/2021/02/01/sparky-food-dryer-utilizing-agricultural-waste-to-enhance-food-security-and-conserve-the-environment/. Used under CC BY-4.0.

of solving the problem. The Sparky Dryer uses biofuel from the garden, such as leaves, sticks, and other inert organic materials. It transforms the toxins into valuable gases so that it may dehydrate the food while emitting no carbon dioxide. It does not need electricity, has a dehydrating capacity five times greater than solar dryers, ten times greater than open sun dryers, and retains the nutrients in the food. With just 2 kg of biofuel, 50 kilograms of food may be dried in 5 hours, extending its shelf life from days to months (Kamau, 2020). The primary goal of Sparky Dryer (Figure 6.1) is to promote no poverty, no hunger, economical, clean energy that is environmentally friendly, simple to use, and easy to clean.

6.4.1.2 Plasma Separation Whirligig Toy

The inspiration for the idea is from the childhood whirligig toy where a loop of string is inserted through two holes in a button. Take hold of the loop's ends and pull rhythmically. The twine coils and uncoils as the button rotates rapidly. A centrifuge is the mainstay of any medical diagnostics facility. Extracting plasma from whole blood requires checking the concentration of pathogens and parasites in blood, urine, and stool. Standard care diagnostics is impossible in Africa because of the lack of medical facilities and equipment. So, the bioengineers at Stanford University developed an inexpensive, human-powered blood centrifuge that helps diagnose and treat diseases like malaria, HIV, tuberculosis, African sleeping sickness, and many more in off-the-grid areas (Newby, 2017). The low-cost whirligig toy helps the centrifugation to separate plasma from blood and other parasites. The market-available centrifuge is expensive, bulky, and requires electricity. In contrast, this ultra-low design centrifuge (in Figure 6.2) is lightweight (2 g), low costs (<20 cents), human-powered, does not require electricity and is made of paper.

6.4.1.3 Mitticool Fridge

Mitticool fridge (Figure 6.3) is a type of refrigerator that is made of clay. Mansukh Prajapati, a potter from Gujarat, India, manufactures it (Mitticool, 2023). He got inspiration from the 2001 earthquake in India; everyone lost their home. The Mitticool fridge is designed to keep food and beverages cool without electricity (Logical Indian,

FIGURE 6.2 Low-cost paper centrifuge.

Source: Paperfuge (2017). https://theindexproject.org/post/paperfuge-2017-play-learning-winner. Used under CC BY-4.0.

FIGURE 6.3 Mitticool fridge by Mansukh Prajapati.

Source: https://thelogicalindian.com/story-feed/get-inspired/high-school-dropout-designed-a-refrigerator-that-runs-without-electricity/. Used under CC BY-4.0.

2015). It works on the principle of evaporative cooling. The outer layer of the fridge is made of porous clay, which allows water to seep through and evaporate. This process cools the fridge's interior, which can then be used to store food and drinks. It is an eco-friendly alternative to traditional refrigerators, as it does not require any electricity to operate. Therefore, it is a unique and innovative product that offers an eco-friendly and affordable solution for keeping food and drinks cool in certain situations. This jugaad innovation fulfills the goal of social, ecological, and economic sustainability.

6.4.1.4 Bamboo Windmill

Wind turbines are frequently seen as an economical, clean, and environment-friendly source of electricity. The blades used in wind turbines are often built of pricey carbon composite materials. As a result, Indian researchers are looking at the viability of employing bamboo for blades, which is inexpensive (BambooGrave, 2023). Efficacy will undoubtedly be lower than other methods, but small-scale installations can still benefit. Bamboos are readily available, easily cultivated locally, lightweight, and quickly regenerable. Supplying renewable energy to rural areas will be an excellent choice since smaller wind farm installations will be adequate there. Due to the abundance of bamboo in China, where eco-friendly products like furniture and flooring are already made, this will also be a practical alternative there. Chinese researchers also promote the bamboo windmill (Figure 6.4) as an appealing alternative for preserving electricity for future generations.

FIGURE 6.4 Bamboo Windmill.

Source: Hines, M. (2010). Bamboo windmills. https://www.trendhunter.com/trends/gijsbert-koren. Used under CC BY-4.0.

6.5 BENEFITS OF FRUGAL PRINCIPLES IN SUSTAINABLE DESIGNING AND MANUFACTURING

Sustainable manufacturing focuses on the product's design, production, manufacture, and inputs. It focuses on creating less hazardous objects that require less energy, produce less trash, are recyclable, and can be reused (Introduction to Sustainable Manufacturing, n.d.). It includes everything from significant expenditures in new technologies and product innovation to minor process changes. This discusses how frugal strategies can be applied to reduce consumption, reuse, recycle, and conserve resources at the design and manufacturing stage. The elementary need is to bridge the strong relationship between designer and user by including the principles of transparency and active engagement principles. The more transparency in the functional part of the product, the more the user can connect themselves with the artifact as it is easy to repair and reuse the things. Also, frugality explores the possibility of shared responsibility of the user and designer in responsibly making the product. Postphenomenological philosophers such as Verbeek discuss how including the user at the design stage can help overcome multiple problems at the very beginning (Verbeek, 2005). The user is the one who will use the product, so by inculcating the role of the user; the designer can anticipate various possibilities of how product functioning can go wrong. Verbeek (2011) outlines the three essential methods that designers can use to maintain their relationship with users through their artifacts: moral imagination, informed prediction, and scenario technique. In moral imagination, the creator imagines how the item will function in the user's life. The method of informed prediction, also known as augmented constructive evaluation, entails involving all the stakeholders in the design and consideration of its social impact, including users, designers, organizations, and others. The virtual presentation of the product serves as the foundation for the scenario and simulation process. Instead of just emphasizing functionality, the product's usefulness in relation to other cultures is now more of a priority. The idea is to make a strong relationship between the designer and the user. Frugality is the best way to implement such principles because the user is a designer and a user simultaneously, as in the example of Mitticool, where the person designs (user at the same time) the fridge to mitigate his near one's primary need. In Frugal, all stakeholders are already involved, so it is easy to anticipate future outcomes. Frugal has various principles that will be effective for designers in manufacturing the products.

6.5.1 BOTTOM-UP APPROACH

Frugal innovation can be divided into top-down and bottom-up categories (Bhaduri and Talat, 2020). Top-down innovation addresses the evolving nature of technical advancements, the post-innovation crises, and the need to create more affordable, less complex, yet compelling products to reach the bottom of the pyramid. However, it is considered ineffective because it does not address people's fundamental problems of what is required. Bottom-up, frugal people employ their innovations directly. Bottom-up frugal innovations are built on the three pillars of transparency, interaction, and mutual responsiveness. Social links, local knowledge, and community use influence bottom-up approaches. According to Bhaduri and Kumar (2011), these inventions fill the gaps created by the market and the government, geared at achieving the objectives of inclusivity and sustainability (Bhaduri and Talat, 2020). Frugal innovations remain far from the radar of policymakers and social elites. The idea is to create a new market for the people and provide resources to the lower strata of society. Jugaad innovators must be clever, deriving economies of scale and scope in whatever they do. An example is the Logan, a no-frills car priced at 5,000 euros, where the research and development team was inspired by frugal technologies and decided to use them to produce more with less (Radjou et al., 2012). In India, for instance, Ratan Tata, the chairman of the Tata Group, launched the Nano in 2009 for just $2,000 (Gaur and Sahdev, 2015). Carlos Ghosn, chairman and CEO of the Renault-Nissan Alliance, once said, "We don't go to the emerging markets to just bring back a product, but to learn something-look new processes or a whole new mindset" (Radjou et al., 2012).

6.5.2 Longevity and Psychological Connection

Only attaining sustainability in production, consumption, and waste is insufficient because the problem is more profound. Consumers discard most things before they wear out. More must be done to address this issue than reducing pollution at various product lifestyle stages. Therefore, Dutch industrial designers founded Eternally Yours to concentrate on product lifespan (Van Hinte, 1997). They look for ways to strengthen the bond between the individual and the product. Their goal is to engage in unconventional design. Therefore, the designers made products people may cherish and utilize for extended periods. Italian designer Ezio Manzini stated, "It is time for a new generation of products that can age slowly and in a dignified way, that can become our partners in life and support our memories." Eternally Yours illustrates three qualities in the life span of products: technical, economic, and psychological. According to them, sustainable development may achieve goals one and two by emphasizing repair and recycling. The psychological lifespan is the primary component. They address the issue of how to maintain this psychological bond between a person and a product. They employ materials that do not deteriorate or age over time. Sigrid Smits creates a fascinating couch. The design on the couch was initially unseen, but as time goes on, it gradually becomes visible and looks even more lovely. As a result, it encourages users and designers to move toward reducing resource waste.

6.5.3 Recycling and Upcycling Intention of Waste Management

Wang et al. (2023) discuss the positive relation between frugality and the recycling intention of waste management. He talks about three characteristics of a frugal person: First, they are self-restrained and prioritize long-term gains; second, they make every effort to maximize the asset's value; and third, social repercussions don't have as much impact on them as they do on other consumers. Some scholars believe that the reason for frugality is always self-motivation and altruistic behavior (Botetzagias et al., 2015). Such frugal consumers seek the maximum utility of products through repair and reuse (Albinsson et al., 2010). It shows consumers with a highly frugal mindset are highly involved in recycling material, thus contributing to the vital goal of sustainability. Recycling is when we use materials repeatedly, while upcycling is when we transform those resources into something brand-new, functional, and innovative (Waste to Wonder, 2016). The Rock Garden in Chandigarh, India, is a well-known example of upcycling, where garbage symbolizes creativity and inherent worth (Ramaswamy, 2015). The waste (broken bottles, wires, sockets, charred bricks, street lights, bicycle handles, etc.) was elegantly and imaginatively converted into art by Nek Chand, showing how even damaged and unusable items have meaning (Figures 6.5 and 6.6). Laotian culture serves as another illustration. Laos was frequently bombarded during the American-Vietnamese War, and cluster bombs were also dropped there (Khng, 2022). Still, many of them remain unexploded, and fatalities are occurring. The Laotian populace is disarming those bombs and repurposing the material to make tools, jewelry, and other household

FIGURE 6.5 Sculptures from mud and shattered bangles.

Source: Statues made of waste Bangles at Rock Garden Chandigarh. https://commons.wikimedia.org/wiki/File:Statues_made_of_waste_Bangles_at_Rock_Garden,_Chandigarh.jpg, Retrieved on 25 June 2024. Used under CC BY-4.0.

FIGURE 6.6 Sculptures crafted using repurposed glass and ceramic materials.

Source: The Waste Maker. (2015). *The Indian Express*. https://indianexpress.com/article/lifestyle/the-waste-maker/. Used under CC BY-4.0.

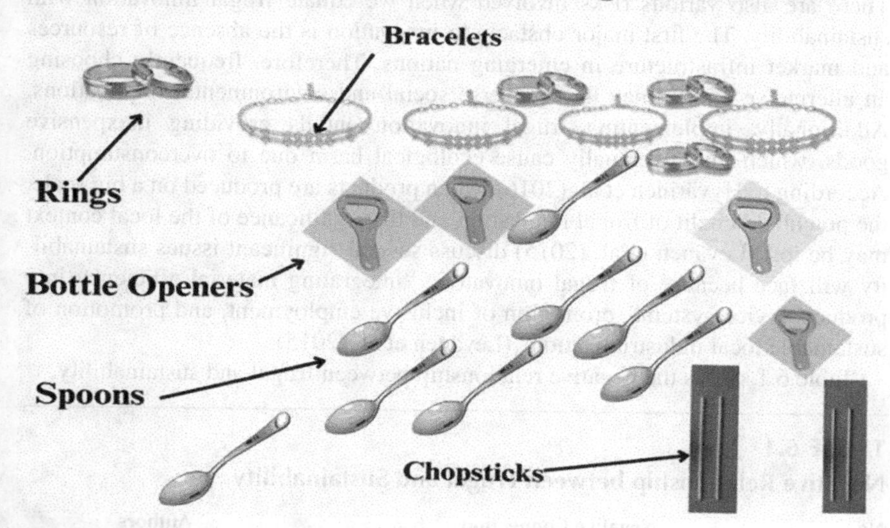

FIGURE 6.7 Upcycling bombs into jewelry, spoons, etc.

Source: Out of courtesy (author's work).

items (Figure 6.7) (Cooper, 2018). The bombs are no longer a disaster for them; they are necessary for survival. So, creating things through economic means is a creative and engaging process. So, upcycling through frugal ways is an innovative and most engaged way of manufacturing products (Figure 6.8).

FIGURE 6.8 Bomb Boats.

Source: Cortesi, M. (2017). Tha Bak Bomb Boats. https://www.atlasobscura.com/places/tha-bak-bomb-boats. Used under CC BY-4.0.

6.6 NEGATIVE RELATION BETWEEN FRUGALITY AND SUSTAINABILITY

There are also various risks involved when we equate frugal innovation with sustainability. The first major obstacle to innovation is the absence of resources and market infrastructure in emerging nations. Therefore, frequently choosing an alternative option may have adverse social and environmental implications. Additionally, implementing frugal innovation entails providing inexpensive goods, which may eventually cause ecological harm due to overconsumption. According to Hyvärinen et al. (2016), when products are produced on a big scale, the potential benefit of frugal innovation and the significance of the local context may be lost. Levänen et al. (2015) discuss several significant issues sustainability will face because of frugal innovators: "integrating material efficiency into product/service systems, promotion of inclusive employment, and promotion of sustainable local industrialization" (Levänen et al. (2015).

Table 6.1 shows the negative relationship between frugal and sustainability.

TABLE 6.1
Negative Relationship between Frugal and Sustainability

No.	Negative Connection	Authors
1	The economic and ecological impact of frugal is not clear.	Rosca et al. (2017)
2	Frugal innovation is not always green innovation.	Sharma and Iyer (2012)
3	The side effect of the impulsive frugal initiative on environmental sustainability.	Wohlfart et al. (2016)
4	Sustainability is not the first focus for frugal innovation.	Weyrauch and Herstatt. (2016)
5	At the conceptual level, it is problematic to equate frugality and sustainability.	Levänen et al. (2015)

6.7 CONCLUSION

Therefore, frugality and sustainability are two interconnected concepts that are becoming increasingly relevant in today's world. Frugality refers to using resources efficiently, minimizing waste and consumption, and focusing on simplicity. On the other hand, sustainability is the ability to satisfy the present generation's demands without compromising future generations' needs. The principles of frugality, such as reusing and repurposing, can contribute to sustainability by reducing the need for new resources and reducing waste. In addition, sustainable practices, such as using renewable energy sources and reducing greenhouse gas emissions, can help preserve resources and create a more frugal future. Moreover, frugality and sustainability principles can contribute to social justice by ensuring that resources are distributed fairly and not wasted unnecessarily. It also encourages the development of innovative and affordable solutions that accommodate the needs of the lower sections of society.

While some scholars have highlighted a negative relationship between frugality and sustainability, such as the risk of overconsumption due to resource availability, frugal is not sustainable; this perspective is one-sided. The issue of overconsumption is not solely a result of resource availability but also resource accumulation. In earlier times, frugal means were used as a principle of virtue, and people lived a more contented life with sufficient resources. Moreover, it is a misconception that frugal innovation is not sustainable or does not consider environmental and social concerns. Mainstream industries have adopted frugal practices, resulting in innovative and inclusive products, as demonstrated by Renault-Nissan and the Nano Car. Frugality is about unique and creative ideas that also consider ethical implications. Overall, frugality and sustainability are essential concepts that can contribute to a more equitable and sustainable future. Their integration into daily practices and manufacturing processes can result in more efficient, cost-effective, and environmentally friendly outcomes in terms of resource utilization.

REFERENCES

Akbar, S., & Subramaniam, N. (2019). Linking frugal innovation and sustainable development: Leveraging corporate accountability mechanisms. In *Frugal Innovation*, 196–211. Routledge. https://doi.org/10.4324/9780429025679-12

Albert, M. (2019). Sustainable frugal innovation—the connection between frugal innovation and sustainability. *Journal of Cleaner Production*, 237: 117747.

Albinsson, P. A., Wolf, M., & Kopf, D. A. (2010). Anti-consumption in East Germany: Consumer resistance to hyper consumption. *Journal of Consumer Behaviour*, *9*(6): 412–425. https://doi.org/10.1002/cb.333

Antonio, A. (2010). Frugality. *IESE Business School Working Paper*, 873. https://doi.org/10.2139/ssrn.1701699

Apio, M. (2021). Eco-friendly food dryer reduces food waste, improves farmers' incomes in Uganda. https://realiti6.wordpress.com/2021/02/01/sparky-food-dryer-utilizing-agricultural-waste-to-enhance-food-security-and-conserve-the-environment/

Arrhenius, S. (1896). On the influence of carbonic acid in the air upon the temperature of the ground. *The London, Edinburgh, and Dublin Philosophical Magazine and Journal of Science*, *41*(251), 237–276.

BambooGrave. (2023). Bamboo as a material for Wind Turbine Blades. https://www.bamboogrove.com/bamboo-turbine-blades.html

Bhaduri, S. & Talat, N. (2020). RRI beyond its comfort zone: initiating a dialogue with frugal innovation by 'the vulnerable'. *Science, Technology and Society*, *25*(2), 273–290.

Bhaduri, S., & Kumar, H. (2011). Extrinsic and intrinsic motivations to innovate: Tracing the motivation of 'grassroot' innovators in India. *Mind & Society*, *10*(1), 27–55.

Botetzagias, I., Dima, A. F. & Malesios, C. (2015). Extending the theory of planned behavior in the context of recycling: the role of moral norms and of demographic predictors. *Resources, Conservation and Recycling*, 95: 58–67, https://doi.org/10.1016/j.resconrec.2014.12.004

British Council. (2018). Refrigeration- 'Sparky Dryer' and 'Cold Hubs' mean cold food in Hot Sun. https://www.britishcouncil.com.sn/sites/default/files/18_refrigeration_-_sparky_dryer_a2_v3.pdf

Cooper, H. (2018). In Laos: People are recycling bombs from the Vietnamese war into jewelry. https://www.vice.com/en/article/3k574v/in-laos-people-are-recycling-bombs-from-the-vietnam-war-into-jewelry

Cortesi, M. (2017). Tha Bak Bomb Boats. https://www.atlasobscura.com/places/tha-bak-bomb-boats

Ehrlich, P. R., & Ehrlich, A. H. (2009). The population bomb revisited. *The Electronic Journal of Sustainable Development*, *1*(3), 63–71.

Frugal. (2020). Online Etymology Dictionary. https://www.etymonline.com/word/frugal

Gaur, L., & Sahdev, S. L. (2015). Frugal innovation in India: The case of Tata Nano. *International Journal of Applied Engineering Research*, *10*(7): 17411–17420.

Genus, A., & Stirling, A. (2018). Collingridge and the dilemma of control: Towards responsible and accountable innovation. *Research Policy*, *47*(1): 61–69.

Ghosh, S. (2020). How rhizophora mangroves on car nicobar islands fought back a rapid sea-level rise on 2004 tsunami. https://india.mongabay.com/2020/08/how-rhizophora-mangroves-on-car-nicobar-islands-fought-back-a-rapid-sea-level-rise-in-2004-tsunami/

Hines, M. (2010). Bamboo windmills. https://www.trendhunter.com/trends/gijsbert-koren

Hyvärinen, A., Keskinen, M., & Varis, O. (2016). Potential and Pitfalls of frugal innovation in the Water Sector: Insights from Tanzania to Global Value Chains. *Sustainability*, *8*(888): 1–16.

Imperatives, S. (1987). Report of the World Commission on Environment and Development: Our common future. 1-300.

Introduction to Sustainable Manufacturing. http://www.gcpcenvis.nic.in/PDF/introduction_to_sustainable_manufacturing.pdf

Kamau, T. (2020). Sparky Dryer. *Engineering for Change*. https://www.engineeringforchange.org/solutions/product/sparky-dryer/

Khng, D. (2022). Laos' war souvenirs: Jewellery, cutlery from unexploded bombs. http://www.weeworks.wkwsci.ntu.edu.sg/WeeVolunteer/stories/2018/war-souvenirs.html

Leliveld, A., & Knorringa, P. (2018). Frugal innovation and development research. *The European Journal of Development Research*, *30*(1): 1–16.

Levänen, J., Hossain, M., Lyytinen, T., Hyvärinen, A., Numminen, S., & Halme. M. (2015). Implications of frugal innovations on Sustainable Development: Evaluating Water and Energy Innovations. *Sustainability*, *8*(4): 1–17.

Logical Indian. (2015). High school dropout designed a refrigerator that runs without electricity. https://thelogicalindian.com/story-feed/get-inspired/high-school-dropout-designed-a-refrigerator-that-runs-without-electricity/

Meadows, D. H., Meadows, D. H., Randers, J., & Behrens, W. W. (1972). The limits to growth: a report to the club of Rome's project on the predicament of mankind, New York: Universe Books, 91 (2).

Mitticool. (2023). Mansukh Prajapati. https://mitticool.com/mansukhbhai-prajapati/

Nash, J. A. (1995). Towards the Revival and Reform of the Subversive Virtue: Frugality. *The Annual of the Society of Christian Ethics*, 15: 137–160.

Newby, K. (2017). Inspired by a whirligig toy Stanford bioengineers develop a 20-cent, hand-powered blood centrifuge. https://news.stanford.edu/2017/01/10/whirligig-toy-bioengineers-develop-20-cent-hand-powered-blood-centrifuge/

Paperfuge. (2017). The Index Project. https://theindexproject.org/post/paperfuge-2017-play-learning-winner

Radjou, N., Prabhu, J., & Ahuja, S. (2012). *Jugaad innovation: Think frugal, be flexible, generate breakthrough growth*. John Wiley & Sons.

Ramaswamy, C. (2015). Building a fantasy from urban waste. https://www.thehindu.com/features/homes-and-gardens/building-a-fantasy-from-urban-waste/article7485421.eces

Rosca, E. M., Arnold, J. C. & Bendul. (2017). Business models for sustainable innovation - an empirical analysis of frugal products and services. *Journal of Cleaner Production*, 162: 133–145.

Sarokin, D. (2022). A brief history of sustainability. In *Corporate sustainability: Does it make a difference?*, 979-8419287587

Schumacher, E. F. (1973). *Small is beautiful: Economics as if people mattered*. Harper & Row Publishers.

17 Goals. Department of Economic and Social Affairs: Sustainable Development. https://sdgs.un.org/goals

Sharma, A., & Iyer, G. R. (2012). Resource-constrained product development: Implications for green marketing and green supply chains. *Industrial Marketing Management*, *41*: 599–608.

Simoes, L., Garrdo, S. R. & Carvalho, A. (2018). Assessing the social sustainability of frugal products. *Social LCA*, *86*, pp. 1–10.

Sustainability. (2023). Wikipedia. https://en.wikipedia.org/wiki/Sustainability. Last edited on 19, June, 2024.

Sustainable. (2022). Online Etymology Dictionary. https://www.etymonline.com/word/sustainable

Van Hinte, E. (1997). *Eternally Yours: visions on product endurance*. 010 Publishers.

Verbeek, P. P. (2011). *Moralizing technology: Understanding and designing the morality of things*. University of Chicago press.

Verbeek, P. P. (2005). *What things do: Philosophical reflections on technology, agency, and design*. Penn State Press.

Wallace-Wells, D. (2019). Jared Diamond: There's a 49 percent chance the world as we know it will end by 2050. https://nymag.com/intelligencer/2019/05/jared-diamond-on-his-new-book-upheaval.html

Wang, H., Bai, R., Zhao, H., Hu, Z., & Li, Y. (2023). Why does frugality influence the recycling intention of waste materials? *Front Psychology*, *13*: 1–8, 2023. https://doi.org/10.3389/fpsyg.2022.952010

Waste Maker. (2015). The Indian Express. https://indianexpress.com/article/lifestyle/the-waste-maker/

Waste to Wonder. (2016). The Hindu. https://www.thehindu.com/features/kids/Waste-to-wonder/article14475868.ece

Weyrauch, T. C. & Herstatt. (2016). What is frugal innovation? Three defining criteria. *Journal of Frugal Innovation*, *2*(1): 1–17.

Wohlfart, L., Bünger, M., Lang-Koetz, C., & Wagner, F. (2016). Corporate and grassroot frugal innovation: A comparison of top-down and bottom-up strategies. *Technology Innovation Management Review*, *6*(4): 5–17.

7 Environmentally Responsible Product Design and Model Making

Rupesh Surwade, Kanwaljit Singh Khas, and Abhishek Bangre
Lovely Professional University, India

7.1 INTRODUCTION

As society advances, people have grown more conscious of the harmful effects of using resources and energy on the surroundings and the community. As an outcome, it is important to build an environmentally friendly development standard that links society, the economy, and the environment. Although it is a crucial aspect of humanity, applying energy design arises by starting sustainable development (Dasgupta, 2007) centered around the concept of sustainability; it mixes conventional and eco-design components to produce an innovative sustainable design (Anastas & Zimmerman, 2007). To establish environmental, social, and overall economic equilibrium, as well as to enhance the ideal society's continuing and coordinated growth, it is not damaging to future generations to fulfill their requests while simultaneously serving the requirements of the present. Sustainable design considers all aspects of human existence, including occupation, manufacturing, energy, urbanization, mobility, interaction, and economy. These have been modified to make up for the negative effects that the development of technologies and science has had as a result of this way of thinking (Clark, Kosoris, Hong, & Crul, 2009). Given the increasing demand for products, their sustainability is more crucial than ever. One tactic to reduce a product's detrimental impact on society, the economy, and the surroundings is to include sustainability in all phases of a product's design process. Sustainability, therefore, addresses social, environmental, and economic issues. Sustainable product design considers environmental and social aspects from the earliest stages of the product creation process to prevent adverse environmental consequences during the life cycle course of the item, as well as comply with the targets for sustainable development.

DOI: 10.1201/9781003309123-7

For the design of sustainable products, consideration of the economy, society, and the environment is crucial (Ashby, Shercliff, & Cebon, 2018). When everything is generally thoroughly planned at a time of massive manufacturing, a plan represents an extraordinary tool by which humans shape the world we live in. This tool's application to the governance of the environment is undeniably broad. After reviewing the key concepts of environmental sustainability in industry, it is feasible to think about how these concepts may be incorporated into the creation of contemporary cycles and products. As soon as the needs that motivate configuration are distilled into examples as well as streams of related concepts, taking in the requirements of the developing ecosphere, configuration ultimately transforms through a cycle that evolves. The integration is primarily composed of trim energy as well as resource streams for such purposes as meeting the requirements of humanity. In its early phases, controlled improvement evolved into the idea of a practical plan, which is today a crucial component of humanity's response to planetary environmental changes. It creates a new financial strategy based on the idea of sustainability by fusing conventional and eco-plans. If individuals can fulfill their demands while simultaneously taking care of the requirements of the present, attaining peace in the social, economic, and environmental spheres, and sustaining and contributing to the advancement of human civilization as a whole, it won't damage them in the future. A realistic plan takes into account all areas of human existence, including labor, production, energy, development, transportation, connection, and the whole economy (Caïd, 2006).

7.2 OBJECTIVES

To educate students, manufacturers, and professionals by demonstrating how to include the many energy sources including solar, wind, or bioenergy.

To create a model or prototype to comprehend and appreciate the necessity for environmentally friendly and energy-efficient goods.

To recognize the many obstacles that the marketing of energy goods faces.

7.3 ENVIRONMENT-FRIENDLY DESIGN

Designing with sustainability means rethinking how to accommodate expansion while minimizing harmful effects on the environment and human health. Technology advancements are not always the focus. It is essential to understand the idea of "decoupling," which seeks to sever the connection between economic growth and environmental degradation (Ehrenfeld, 2008). Because of this, it was able to have a comprehensive grasp of the consequences of environmental problems on the design trend at the beginning of the 1990s, encompassing the most different zones and represented by the consequences of the key encounters (Finkbeiner, Schau, Lehmann, & Traverso, 2010). Following these discussions, there was a time of greater awareness of fresh asset protection standards, which incorporated an assortment of innovative concepts generated without the application of conventional design methodologies that coordinated environmental requests. Although it wasn't always obvious, the notion of "Design to Benefit the Environment" has advanced over the past ten years. It started in a reductive way as a design concept meant to cut down on contemporary waste and increase material application, and as a consequence, it got a better measurement (Hallgrimsson, 2012).

7.3.1 Environmental and Energy-Conscious Product Design

A design idea known as "sustainable design" strives to enhance the quality of consumer goods and lifestyle settings while minimizing or eliminating harmful environmental effects. Finding the best solution that balances sustainability concerns and effectiveness, comfort, aesthetics, economics, and a range of other design issues is another goal of sustainable design (Kristensen, 2004). The environmentally friendly design may be an automatic process for talented designers who have successfully incorporated the principles into the way they design, even though it involves intention.

A sustainable design approach considers how architectural choices impact the environment and building inhabitants, especially the bottom side. A majority of the time, prioritizing environmentally friendly design does not entail abandoning other program requirements like time and budget. Instead, think of ecologically conscious design as a set of criteria to take into account as you consider design choices (Kumar, Singh, & Khurana, 2012, December).

A method of reducing the environmental effect of items is through a product design that benefits the surroundings. Large-scale raw material use and polluting processes may hurt the environment. According to the vast quantity of energy used and the difficulties experienced during disposal, the effect may also be adverse. Therefore, it is necessary to consider a product's whole life cycle, which is from creation through disposal. Whenever developing products, all three of the essential requirements must be given consideration (Dasgupta, 2007). Design for Environment (DfE) is a methodical evaluation of design efficacy in connection to the environment, good health, as well as safeguarding, as well as goals for sustainability, across each stage of the production cycle.

7.3.2 The Idea of Sustainability Is as Follows

The concept of sustainable design is gaining popularity over the past 20 years. It is a concept that acknowledges that civilization as a whole is a part of the environment and that the preservation and preservation of nature are necessary for the survival of human society (Robert, Parris, & Leiserowitz, 2005). Through inventions that illustrate conservation concepts and encourage their adoption in our daily lives, the environmentally friendly design communicates this idea.

A complementary notion that supports environmentally friendly design is typically the belief that throughout time, life is generated and sustained upon an actual collective basis while the majority of these different groups (bioregions) feature mutually supporting life systems that are normally self-sustaining. The principle of sustainable design holds that developing technology must primarily function inside bioregional designs and sizes. We must protect biological diversity and sustainability, and improve the state of the atmosphere, water, and soils, taking into account bioregional feasibility and minimizing the effects of human activity (Surwade, Khas, Raghani, & Kamal, 2023a).

Whether it is called "sustainable design," for instance, "sustainable products the universe," "designing for nature," "environmentally conscious and sensitive design," or "holistic asset administration," sustainability refers to the capacity of earth's and human's cultural systems to be reproduced into almost everything.

Sustainability requires a decrease in the quality of life, yet it does require a mentality change, a shift away from consumerist values. These advancements must take into account social responsibility, environmental conservation, global interconnection, and economic viability.

7.3.3 To Accomplish Sustainability, the Following Design Principles Have to Be Adhered To

- Uphold the right of nature and people to exist in a way that is long-lasting, wholesome, diverse, and healthy.
- Appreciate how crucial interdependence is. The natural environment interacts with and depends on elements of human design, which has broad implications for all sizes. Include even distant effects in your design decisions.
- Appreciate how spirit and matter interact. Think about the connections between spiritual and material consciousness that exist and are developing in many elements of human settlement, such as society, housing, business, and industry.
- Assume responsibility for how design decisions affect coexistence rights, environmental sustainability, and the health of humans.
- Produce durable, secure goods. Do not require upkeep or cautious management of potential risks resulting from careless creation of commodities, procedures, or standards.
- Do away with the idea of waste. To attain the waste-free state of systems that are natural, analyze and improve a product's entire lifecycle.
- Rely upon the energy stream from nature. Like the natural world, human inventions depend on an endless source of sun energy for their creative drive. Incorporate this power practically and effectively.
- Be aware of design limitations. No human creation lasts constantly, and design may not always solve problems. Those who organize and make things should be modest in the midst of nature. Do not view the environment as a nuisance that should be avoided or tamed but rather as a teacher and mentor.
- Continue to get better through sharing knowledge. To balance long-term environmental concerns with moral responsibility and restore the natural connection between humanity and environmental technology, promote direct and open communication between partners, customers, manufacturers, and users.

1. Customer demand.
   Making a product with minimal environmental effects gives you a competitive edge.
2. Governmental pressure.
   i. Reduced environmental effect criteria for products are additionally enforced by federal agencies and departments like the National Centre for Solar Power and the Department of Renewable and Sustainable Energy.
      This kind of oversight will only intensify over time.
   ii. The components that comprise numerous items, including packaging, computers, and automobiles for transportation, must now be recycled.
3. ISO (International Organization for Standardization) requirements
   i. Standards (ISO14001-2015) are being developed to promote the practice of design for the benefit of the environment.
   ii. The commercial dynamics that require design for sustainability as a vital component of contemporary product creation underlie all of those endeavors

7.4 DESIGNING SUSTAINABLE PRODUCTS

Sustainable goods are those that help society, the natural environment, and the financial system throughout their whole lifespan, from the extraction of raw materials to final recycling, while also preserving the general well-being, welfare, and the preservation of the environment. This refers to the process of creating a product using environmentally friendly development, social, and economic concepts. It is the procedure of creating a product in line with ethical standards for all three domains of society (Anastas & Zimmerman, 2007).

Conventional design that incorporates consideration of these behavioral changes must adopt a different strategy from sustainable design. Under this novel approach, each design choice must be weighed against the cultural and ecological heritage of communities, regions, and global settings.

There are no established criteria for designing for sustainability because it is presently in its infancy (Ryszard et al., 2021). There are many challenging trade-offs involved, and there are few unbreakable laws. Engineers can utilize the standards published by several groups to aid in the creation of environmentally friendly designs. These standards usually have lengthy checklists, which include the following:

- Recyclable use materials, prevent materials made of composites and standardize items, especially fasteners.
- Prevents permanent connections of unsuitable materials, such as welds, while making it simpler to remove components.
- Provides conventional layouts for components with many functions.
- Reduce waste by minimizing product size, weight, and container.

- Lessen the quantity of energy consumed by items and that utilized in manufacture.
- Create products and parts with many uses, specify materials that are both renewable and recycled, use remanufactured components, design for excellent performance and longevity, and, finally, plan for integrated cycling.

Design for Sustainability – Design for sustainability is the method of developing tangible and intangible products while taking into account the environment, economics, and social values mentioned in Figure 7.1.

Continuous development and innovation management Cost, aesthetics, usability, and simplicity of manufacturing are just a few of the many competing variables that must be taken into account while creating new goods. (Sadhukhan, Joshi, Shemfe, & Lloyd, 2017) Design for Sustainability and Design for Manufacturing are two examples of important needs that are usually referred to as "design." The awareness of sustainability has increased, so too has a design for the environment, referred to as green design, and designing for environmental responsibility, also recognized as the environmentally friendly product development process.

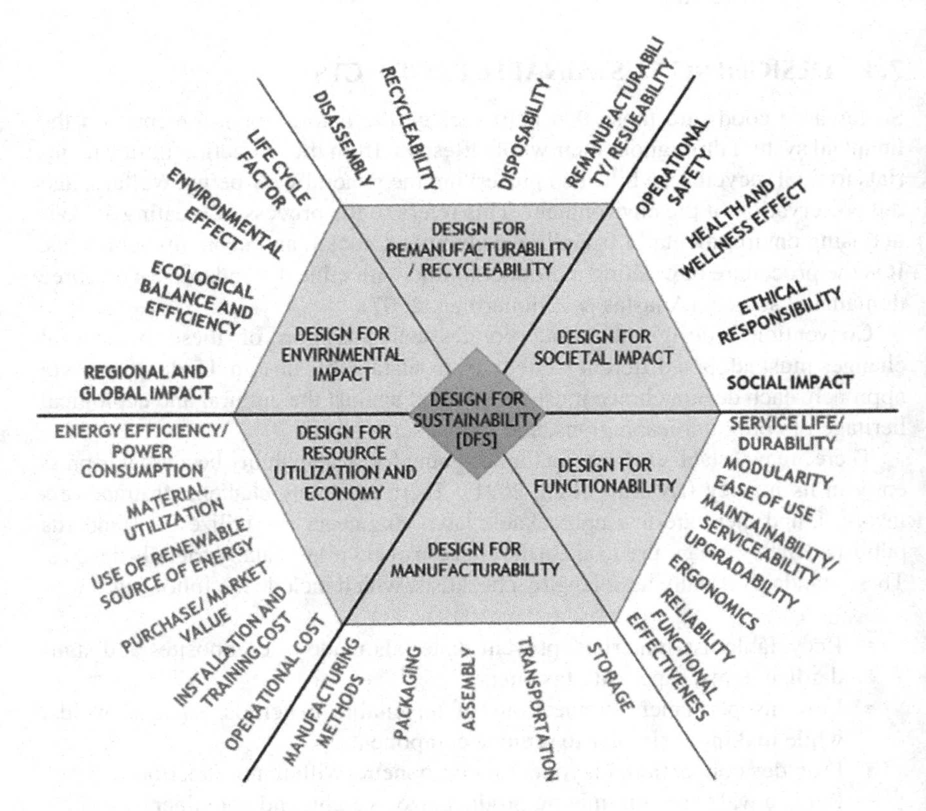

FIGURE 7.1 Important components of "design for sustainability".

7.4.1 Benefits of Environmentally Friendly Goods

Figure 7.2 shows the importance of environmentally friendly products for society.

- **Preserve Energy**: Solar-powered boards are one example of an eco-friendly device that uses energy from the sun. They seem to be a workable substitute for the transmission of energy based on petroleum.
- **Improved health**: Eco-friendly products and green buildings are better for the surroundings since they don't include any dangerous synthetic components or materials (Simon, 1996).
- **Saves Environment**: They save the environment by avoiding the use of fossil fuels. They also aid in lowering carbon dioxide levels in the environment, so preventing environmental change.
- **Material savings**: Environmentally friendly goods use non-harmful components without compromising quality, ensuring efficiency.
- **Conserves water**: A green building contributes to water protection. Additionally, it offers reused water in addition to other water sources like water.
- **Costs Less**: Because green buildings conserve both water and energy, these structures are less expensive. Although construction might seem more expensive initially, it will ultimately result in reduced activity and ongoing costs.

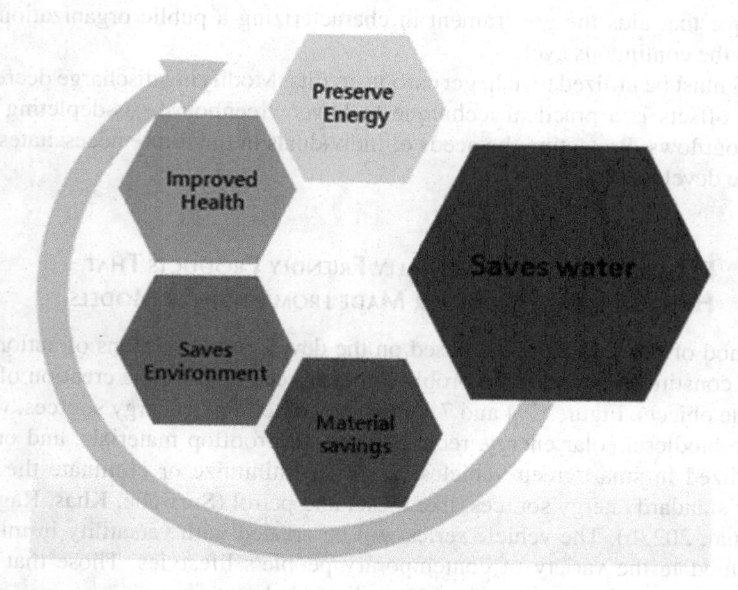

FIGURE 7.2 Benefits of eco-friendly products.

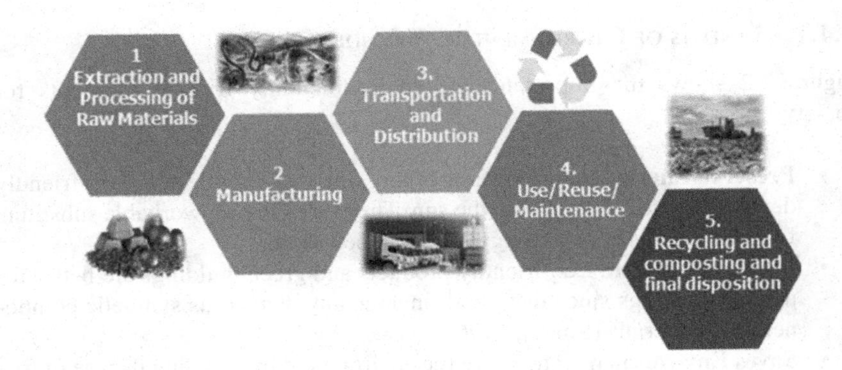

FIGURE 7.3 PDS.

7.4.2 Product Development Stages (PDS)

Figure 7.3 illustrates how the PDS is a technique for analyzing environmental influences at every stage in a product's life cycle. It comprises preparation, manufacturing, appropriation, application, upkeep and repair, and disposal or recycling from the extraction of raw materials through the last processing of the result (Suman, 2023). PDS highlights the advantages of goods to the environment, society, and economy throughout a product's entire existence cycle, starting with the gathering of raw materials. PDS is a tool that examines a product's or implementation's numerous environmental consequences (Eybye, 2014). It is a common technique that aids the government in characterizing a public organization and adds to the continuous cycle.

PDS must be utilized to help get carbon credits. Modifying discharge decreases carbon offsets is a practical technique to lower greenhouse gas-depleting substance outflows. Protecting the needs of individuals in the future necessitates sustainable development.

7.4.3 Examples of Environmentally Friendly Products That Have Been Prototyped or Made from Physical Models

A method of making decisions based on the development of plans of action, the design constitutes a method of problem-solving centered on the creation of new valuable objects. Figures 7.4 and 7.5 show how alternative energy sources, which include biodiesel solar energy, recharging power, rooftop materials, and others, are utilized in smart/green vehicles. This will minimize or eliminate the need for our standard energy sources, like diesel and petrol (Surwade, Khas, Raghani, & Kamal, 2023b). The vehicle series will be created with versatility in mind to accommodate the variety of contemporary people's lifestyles. Those that were completed by students and professors are listed below.

Figure 7.4 A solar-powered folding scooter (powered by a solar panel). Submitted by Krishna Kabbra, M. Des, PIADS Nagpur.

FIGURE 7.4 Solar foldable scooter design by Krishna Kabbra, M. Des, PIADS Nagpur.

Source: Out of courtesy (author's work).

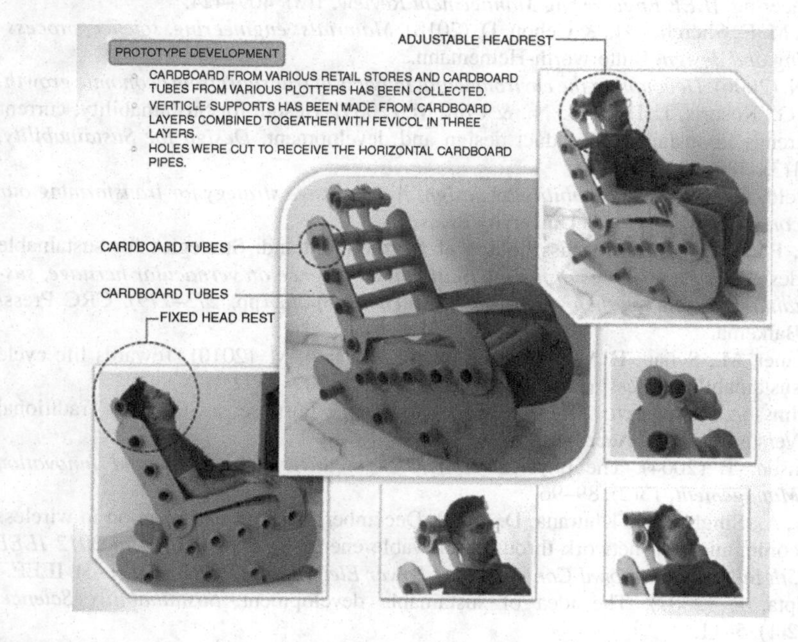

FIGURE 7.5 Corrugated sheets armchair designed by Sarang Holey, M. Des, PIADS. Nagpur.

Source: Out of courtesy (author's work).

7.5 CONCLUSION

The user encounter with environmentally friendly products is now mostly associated with inefficiency or items that allow for material reuse. Protecting people's interests for the future necessitates sustainable development. Additionally, designers are interested in methods to enhance environmental responsibility, improve human conditions, and create new concepts and values. To accomplish environmental sustainability along with energy improvement, widespread mindfulness is necessary. By using life cycle evaluation approaches, product designers are addressing the energy constraints they face and starting to think about their environmental obligations.

The usage of controlled or unconventional products has a less negative effect on the surroundings and can last for a longer time. Minimizing or restricting the consumption of nonrenewable assets, monitoring non-depletable resources, and minimizing or restricting the usage of nonrenewable assets are the objectives of sustainable design. Traditional design approaches should be developed and expanded upon by sustainable design. While we recognize the significant advancements made in the sustainability environmental area, product design and associated practices are also as important.

REFERENCES

Anastas, P. T., & Zimmerman, J. B. (2007). Design through the 12 principles of green engineering. *IEEE Engineering Management Review*, 1(3), 409–424.

Ashby, M. F., Shercliff, H., & Cebon, D. (2018). *Materials: engineering, science, processing and design*. Butterworth-Heinemann.

Caïd, N. (2006). *Decoupling the environmental impacts of transport from economic growth.*

Clark, G., Kosoris, J., Hong, L. N., & Crul, M. (2009). Design for Sustainability: current trends in sustainable product design and development. *Design for Sustainability*, 1(3), 409–424.

Ehrenfeld, J. (2008). *Sustainability by design: A subversive strategy for transforming our consumer culture*. Yale University Press.

Eybye, B. T. (2014). Aisle-truss houses of Northern Jutland: Strategies for sustainable design. *In Proceedings of the international conference on vernacular heritage, sustainability and earthen architecture, Valencia, Spain* (pp. 273–279). CRC Press/Balkema.

Finkbeiner, M., Schau, E. M., Lehmann, A., & Traverso, M. (2010). Towards life cycle sustainability assessment. *Sustainability*, 2(10), 3309–3322.

Hallgrimsson, B. (March, 2012). A Model for Every Purpose: a Study on Traditional Versus. *Idsa.org*. Accessed.

Kristensen, T. (2004). The physical context of creativity. *Creativity and Innovation Management*, 13(2), 89–96.

Kumar, A., Singh, T., & Khurana, D. (2012, December). Energy optimization in wireless communication network through renewable energy sources (RES). *In 2012 IEEE 5th India International Conference on Power Electronics (IICPE)* (pp. 1–5). IEEE.

Dasgupta, P. (2007). The idea of sustainable development. *Sustainability Science*, 2(1), 5–11.

Robert, K. W., Parris, T. M., & Leiserowitz, A. A. (2005). What is sustainable development? Goals, indicators, values, and practice. *Environment: Science and Policy for Sustainable Development*, 47(3), 8–21.

Surwade, R., Khas, K. S., Raghani, S., Kamal, M. A. (2023a). Exploring the potential of model making as a tool for designing sustainable buildings. *Civil Engineering and Architecture*, 11(4), 2231–2239.

Surwade, R., Khas, K. S., Raghani, S., Kamal, M. A. (2023b). Model making as a creative skill and tool for teaching-learning process in architecture and product design. *Civil Engineering and Architecture*, 11(6), 3278–3284.

Ryszard, B., Barbara, S. T., Katarzyna, H. (2021). The idea of sustainable and permanent development in the context of science and business practice. *European Research Studies Journal*, 24(1), 188–200.

Sadhukhan, J., Joshi, N., Shemfe, M., & Lloyd, J. R. (2017). Life cycle assessment of sustainable raw material acquisition for functional magnetite bionanoparticle production. *Journal of Environmental Management*, 199, 116–125.

Simon, H. A. (1996). *The sciences of the artificial*. MIT Press.

Suman, D. & (2023). Sustainable product design for electric vehicles. *In Progress in sustainable manufacturing*, (pp. 31–44). Singapore: Springer Nature Singapore.

8 Polymer-Metal Matrix-Based Composite for Sustainable Manufacturing of Smart Structures

Ankan Shrivastava, Jasgurpreet Singh Chohan, and Ranvijay Kumar
Chandigarh University, Mohali, India

Kamalpreet Sandhu
Lovely Professional University, Phagwara, India

8.1 INTRODUCTION

A range of additive manufacturing (AM) technologies can separately transform data from generated solid models into actual parts. There are several 2D cross-sections with limited widths employed for organizing the information. These cross-sections are sent into AM equipment so that the physical part can be formed, merged, and added layer by layer (Gibson et al., 2021). Many technologies make it possible to use 3D printing, such as stereolithography (SLA), selective laser sintering (SLS), and digital light processing (DLP) etc. (Vardhan et al., 2020). Even though they have several advantages, such as quicker response times and lower costs for the prototype of a new product, current state-of-the-art 3D printing processes can only be used to create objects that cannot change their shape when an external stimulus is applied or a kinematic mechanism is added (Leist et al., 2017). The continuous ongoing development into the 3D printed industry leads to a new technology called 4D printing.

4D printing is another type of AM technique in which a physical component is developed that changes its structure with time when exposed to an outside stimulus (Raviv et al., 2014). Several practical applications, including those in

DOI: 10.1201/9781003309123-8

the fields of aerospace, biomedicine, textiles, and automobiles, have benefited from 4D printing's constantly expanding interest (Zarek et al., 2016). Shape memory polymers (SMPs) provide a special application of 4D printing. A class of materials known as SMPs can maintain a temporary shape while also having the ability to switch to a permanent shape by applying external stimuli like temperature, magnetism, humidity, pH value, etc. (Alshebly et al., 2021). The most common shape memory polymer is polylactic acid (PLA) which is produced by direct condensation of lactic acid and ring-opening polymerization of lactide, which are two processes that can be used to create PLA. It has generated a lot of scientific attention because of its outstanding strength, biodegradability, and biocompatibility, and it is regarded as one of the most successful biomaterials in 3D and 4D printing (Liu and Zhang, 2011). As 4D printing technology develops, the additional way that the use of PLA is growing concerns the many studies that have attempted to combine PLA with various materials to improve its capabilities in 4D printing applications. In this regard, it has investigated the shape recovery characteristics of PLA/C/silicon carbide (SiC) composite filament with various compositions. The result demonstrates that PLA with 50% SiC+10%C has a minimum recovery time of 0.25 seconds to completely recover its original shape (Liu et al., 2018a). Liu et al. designed and produced a Miura-origami sheet with shape memory capabilities using PLA. The warped tessellation object was found to restore to its original shape in approximately 23 seconds at 90°C (Liu et al., 2018b). Lashgari et al. (2016) investigated the shape memory properties of poly (L-lactic acid) (PLLA)/graphene nanoplatelets (GNPs) through infrared light and thermal heating. The result shows that PLLA with 6% of GNPs has the maximum recovery rate of 99.4% by the activation of the stimuli (Lashgari et al., 2016). The shape memory properties of the PLA/wood reinforced with nano silica and nano alumina are performed by Sharma and Singholi. 2021. According to the study, the shape recovery characteristics of wood PLA composite are reduced by 46% and 6.6%, respectively, when nano silica and nano alumina are incorporated (Sharma and Singholi, 2021). The previous study has reported the impact of infill patterns on the shape memory properties of the 3D printed PLA structures (Ehrmann and Ehrmann, 2021). The PLA/TPU shape recovery properties were investigated in the reported study, and it was found out the TPU considerably improved the shape recovery ratio of PLA for the PLA/TPU (50/50) blends, increasing it to 93.5% at 160°C (Lai and Lan, 2013). The impact of various printing conditions on the shape memory capabilities of PLA was examined by Barletta et al. 2021. The activation temperature was the most important factor in causing the return of the initial shape in the shortest amount of time (Barletta et al., 2021). The PLA reinforced with ZnO nanoparticles shape memory capabilities explored by Singh et al. with different FDM process parameters. The result demonstrated that the 99.88% shape recovered at 80% of infill density in 30 minutes when the samples were placed into the water at 70°C (Singh et al., 2022).

Even though there have been many studies on the usage of PLA as SMP, there is still little research conducted on the connection between the 3D printing process's variables and the emergence of an object's shape memory properties. The current study focuses on the uses of PLA/Al composite and how the varying of the FDM process parameters influences the shape recovery capabilities of the PLA composite structures. The shape memory capabilities of the composite have been investigated deeply by varying the various FDM process parameters.

8.2 BACKGROUND OF THE STUDY

The shape memory of the materials is one of the most important considerations for the applications of the 4D printing concept. The Scopus-based database was analyzed to investigate the background of the previous studies. Upon searching the keywords "4D printing" and "PLA" on www.scopus.com, a total of 141 documents were found. The highest number of studies were reported in the year 2022. The trend in Figure 8.1 shows that the studies related to 4D printing of PLA-based materials have increased at a rapid rate after the year 2020.

Figure 8.2 shows the subject-wise publication of the reported research paper related to the 4D printing of PLA materials. It has been observed that engineering and materials science are both the leading subjects, which contributed 29.5% and 28.9%, respectively, followed by computer science, physics, chemistry, etc.

Figure 8.3 shows the map-based text data for the previous studies of the 4D printing. The Vosviewer software package of version 1.6.17 was used to analyze the abstract of the previous studies available on the Scopus website. It has been observed that most of the studies have been reported on the properties of the polymers and their characterization. The less studies have been reported on the action

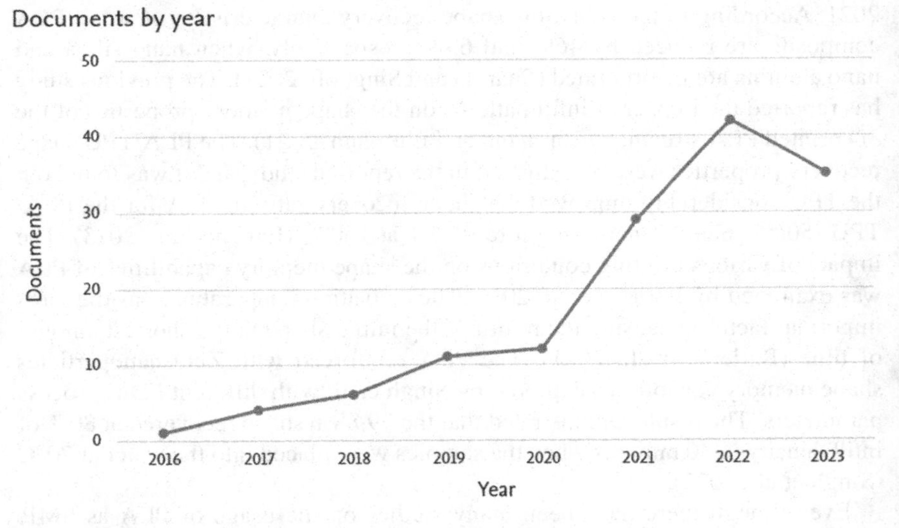

FIGURE 8.1 Year-wise publication related to 4D printing of PLA material.

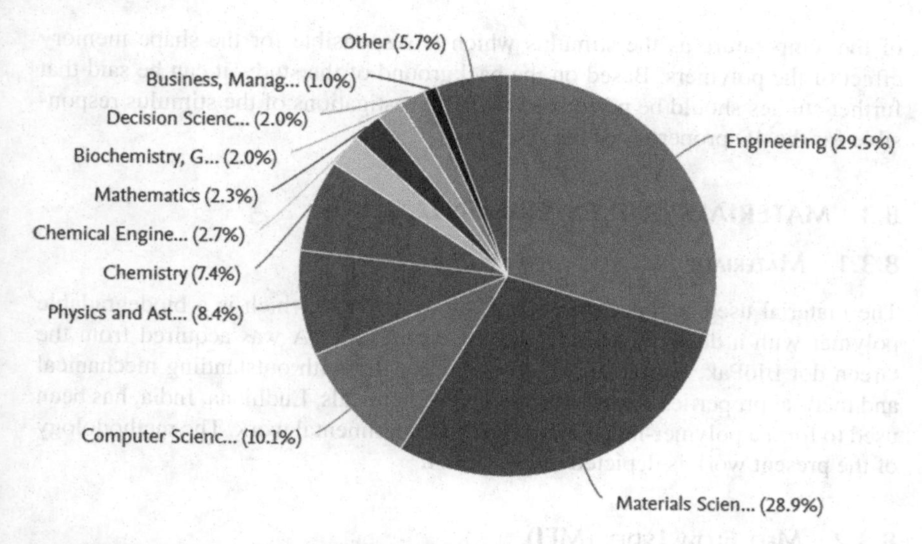

FIGURE 8.2 Subject-wise publication of the research papers.

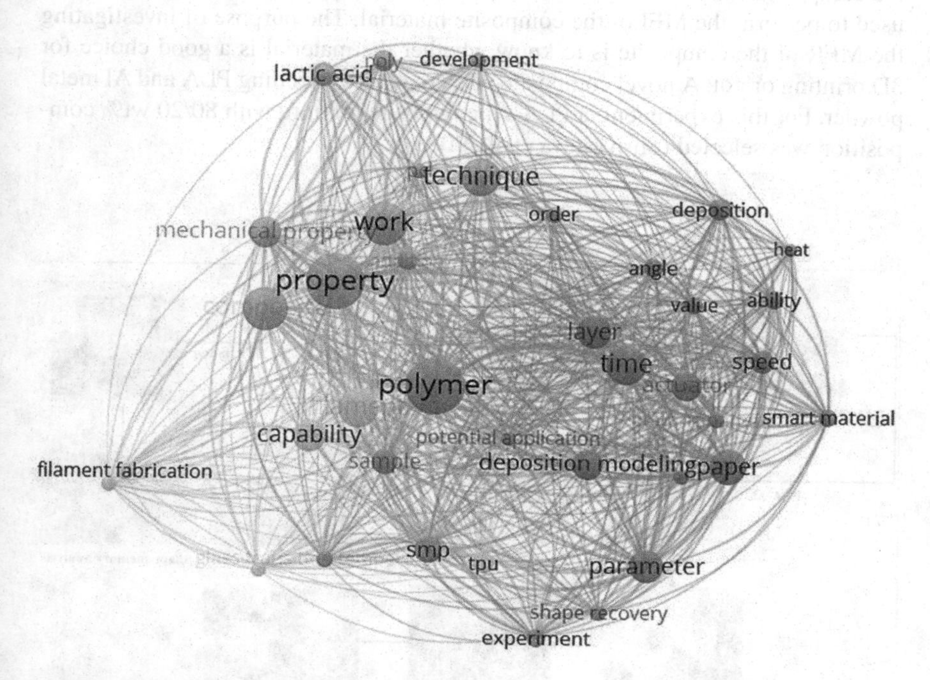

FIGURE 8.3 Map-based text data for publications of 4D printing of PLA material.

of the temperature as the stimulus which is responsible for the shape memory effect of the polymers. Based on the background of the study, it can be said that further studies should be performed on the investigations of the stimulus responsible for the 4D properties of the PLA.

8.3 MATERIALS AND EXPERIMENTATION

8.3.1 Materials

The material used in this experimental study is PLA, which is a biodegradable polymer with a density of 1.24 ± 0.05 g/cm^3. The PLA was acquired from the Green dot BioPak, Gujrat, India. Al metal powder with outstanding mechanical and thermal properties acquired from Shiva Chemicals, Ludhiana, India, has been used to form a polymer-metal matrix in this experimental work. The methodology of the present work is depicted in Figure 8.4.

8.3.2 Melt Flow Index (MFI)

The melt flow index is a flow-testing apparatus that measures the melt flow of the composites. In this study, the melt flow tester (as per ASTM D1238) has been used to perform the MFI of the composite material. The purpose of investigating the MFR of the composite is to know whether the material is a good choice for 3D printing or not. A novel composite was created by blending PLA and Al metal powder. For this experiment, a PLA-Al composite mixture with 80/20 wt% composition was selected (Shrivastava et al., 2023).

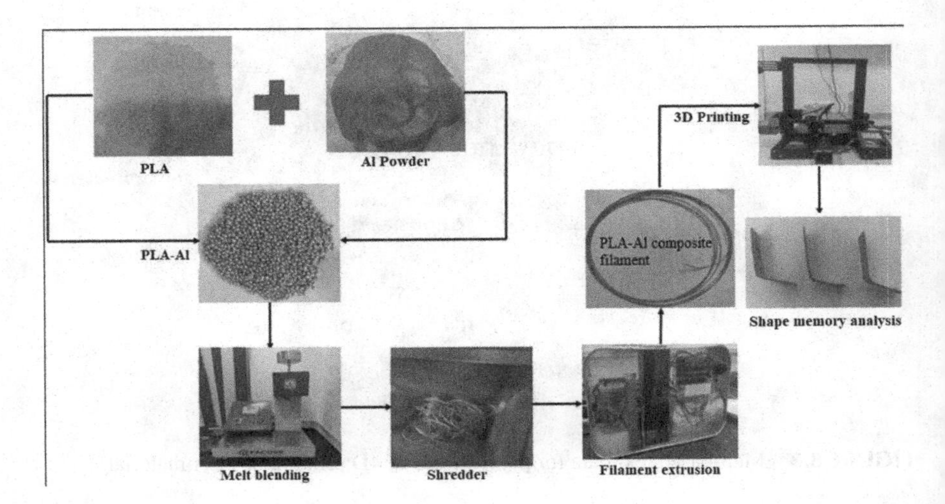

FIGURE 8.4 Procedure for fabrication of the shape memory structures.

8.3.3 Fabrication of Filaments

The filaments that are prepared by the extrusion process consist of 80 wt% of PLA and 20 wt% of Al metal powder. In the extrusion process, the composite in granules form feeds into the extruder through the hopper, and the screw pushes the material through the die in a filament form. The Felfil Evo single screw extruder (make: Felfil, Italy) has been utilized to prepare feedstock filaments for 3D printing at the temperature of 180°C and RPM of 4.

8.3.4 3D Printing of PLA-Al Composite Structures

The Tinker CAD software was used to create the 3D CAD models, which had the following dimensions: 50 mm in length, 10 mm in width, and 1 mm in thickness. The STL file was then converted, and the Ultimaker Cura software was employed to slice the model to generate the final G-code. Nine samples of PLA-Al composite filaments were fabricated to investigate the shape memory properties using a Creality ender pro-3D printer. The printing parameters used in the fabrication of the samples are depicted in Table 8.1. The process parameters and their ranges have been selected based on the observations made in the pilot experimentation.

8.4 RESULTS AND DISCUSSION

First of all, the MFI was ascertained for the PLA-20%Al composites, and it was found suitable for 3D printing. In the next stage, the composite feedstock filament was manufactured using a filament extruder. The shape memory properties of the 3D-printed samples were investigated by performing the bending of the samples. All of the samples were prepared with different printing parameters. All the samples were placed into the pool of water and heated for one hour at 70°C using an oven. Once heated for an hour, the sample was removed from the water and bent to around 90° to examine the shape memory properties. The purpose of

TABLE 8.1
Recovery Angle of All Samples with Time at Different Experimental Conditions

Sample	Layer Thickness (mm)/Infill Percentage (%)/Infill Pattern	Recovery Angle After 1 Hour
1	0.1/20/lines	150°
2	0.1/50/grid	152°
3	0.1/80/zig zag	162°
4	0.12/20/grid	178°
5	0.12/50/zig zag	177°
6	0.12/80/lines	175°
7	0.15/20/zig zag	177°
8	0.15/50/lines	177°
9	0.15/80/grid	177°

heating the samples is to make them flexible so they can be bent without breaking. The samples were bent and then left to cool before being heated up to make them hard so that samples could maintain their temporary shapes. After maintaining their temporary shapes, the samples were again placed in hot water once placed back into the 70°C water pool. Most of the bent samples quickly regain their former shape except samples 1, 2, and 3. Samples 1, 2, and 3, which were prepared using (0.1-layer height, 20%, 50,% and 80% infill density, and infill pattern grid lines zigzag) recovered back to nearly 150° after removing from the hot water in 15 minutes. After placing these samples in hot water for one hour, the samples return to 150°, 160°, and 162° angles, respectively. Samples exhibit good shape recovery properties; they haven't shown any shape recovery after being placed in the heated water for one hour, perhaps because the printing conditions used to manufacture these samples didn't meet the requirements for shape recovery. The other samples, 4, 5, 6, 7, 8, and 9, showed good shape recovery properties. The samples that were taken out of the water after 15 minutes regained nearly 60% of their former shape. When heated to 70°C, samples 4 and 5 were manufactured with a 0.12-layer height, 20, 50 infill density, and grid, zigzag, and infill pattern to expand back to near of previous shape in just 30 minutes, but samples 6 to 9 required one complete hour to return to near of former structure. Sample 4 recovered a better shape of its initial structure, while the other samples didn't regain that much. Some previous studies have been reported on the shape recovery of the materials and their characterization (Qi et al., 2017; Senatov et al., 2016; Zeng et al., 2020; Zhang et al., 2021; Dong et al., 2021; Sorimpuk et al., 2022; Singh et al., 2020; Makwakwa et al., 2022; Ferreira and Andrade, 2021; Liu et al., 2019). Table 8.1 represents the shape recovery of all samples concerning time at different experimental conditions. This indicates that the experiments were effective and could be used to create composites-based shape memory materials by combining them with other metal particles.

This experimental investigation examined the overall shape recovery characteristics of PLA/Al composites. This study demonstrates that the overall shape and condition of a 4D printed product can be printed utilizing 3D printing. The 4D printed component can then be heat treated to turn into the predetermined temporary shape when heated over its glass transition temperature. This temporary shape will be maintained by the 4D printed component at room temperature, but at a temperature over its glass transition temperature, it can be heated to take on its original, permanent shape. Lastly, when heated above its glass transition temperature of 70°C, the PLA/Al composite exhibits excellent shape recovery capabilities that can be applied in many different practical fields, including the biomedical, aerospace, and automotive industries.

8.5 CONCLUSIONS

This study effectively investigates the 4D printing characteristics of the PLA/Al composite 3D printed samples generated by the FDM printer. Samples 1, 2, and 3 have less recovery near the original shape, possibly because the printing conditions utilized to create these samples were responsible for lesser shape recovery, whereas the other sample exhibits impressive shape recovery when heated to 70°C. Sample 4, prepared at 0.12-layer thickness, 20% infill density, and grid infill patterns, exhibited better shape recovery, while samples 1 and 2, which were prepared using 0.1-mm layer height, 20%, and 50% infill density and grid lines infill patterns had a lower shape recovery. According to the study's findings, it can be concluded that the shape can be regained when the material is heated above the glass transition temperature.

ACKNOWLEDGMENT

The authors are thankful to the University Centre of Research and Development, Chandigarh University, for providing the lab facilities and Department of Science and Technology (DST) (File No. SP/YO/2021/2514) for financial assistance.

REFERENCES

Alshebly YS, Nafea M, Ali MS, Almurib HA. Review on recent advances in 4D printing of shape memory polymers. *European Polymer Journal*. 2021;159:110708.

Barletta M, Gisario A, Mehrpouya M. 4D printing of shape memory polylactic acid (PLA) components: Investigating the role of the operational parameters in fused deposition modelling (FDM). *Journal of Manufacturing Processes*. 2021;61:473–80.

Dong K, Ke H, Panahi-Sarmad M, Yang T, Huang X, Xiao X. Mechanical properties and shape memory effect of 4D printed cellular structure composite with a novel continuous fiber-reinforced printing path. *Materials & Design*. 2021;198:109303.

Ehrmann G, Ehrmann A. Investigation of the shape-memory properties of 3D printed PLA structures with different infills. *Polymers*. 2021;13(1):164.

Ferreira WH, Andrade CT. The role of graphene on thermally induced shape memory properties of poly (lactic acid) extruded composites. *Journal of Thermal Analysis and Calorimetry*. 2021;143(4):3107–15.

Gibson I, Rosen DW, Stucker B, Khorasani M, Rosen D, Stucker B, Khorasani M. *Additive manufacturing technologies*. Cham, Switzerland: Springer; 2021.

Lai SM, Lan YC. Shape memory properties of melt-blended polylactic acid (PLA)/thermoplastic polyurethane (TPU) bio-based blends. *Journal of Polymer Research*. 2013;20:1–8.

Lashgari S, Karrabi M, Ghasemi I, Azizi H, Messori M, Paderni K. Shape memory nanocomposite of poly (L-lactic acid)/graphene nanoplatelets triggered by infrared light and thermal heating. *Express Polymer Letters*. 2016;10(4):349–59.

Leist SK, Gao D, Chiou R, Zhou J. Investigating the shape memory properties of 4D printed polylactic acid (PLA) and the concept of 4D printing onto nylon fabrics for the creation of smart textiles. *Virtual and Physical Prototyping*. 2017;12(4):290–300.

Liu H, Zhang J. Research progress in toughening modification of poly (lactic acid). *Journal of polymer science part B: Polymer Physics*. 2011;49(15):1051–83.

Liu W, Wu N, Pochiraju K. Shape recovery characteristics of SiC/C/PLA composite filaments and 3D printed parts. *Composites Part A: Applied Science and Manufacturing*. 2018a;108:1–1.

Liu Y, Zhang F, Leng J, Fu K, Lu XL, Wang L, Cotton C, Sun B, Gu B, Chou TW. Remotely and sequentially controlled actuation of electroactivated carbon nanotube/shape memory polymer composites. *Advanced Materials Technologies*. 2019;4(12):1900600.

Liu Y, Zhang W, Zhang F, Lan X, Leng J, Liu S, Jia X, Cotton C, Sun B, Gu B, Chou TW. Shape memory behavior and recovery force of 4D printed laminated Miura-origami structures subjected to compressive loading. *Composites Part B: Engineering*. 2018b;153:233–42.

Makwakwa D, Motloung MP, Ojijo V, Bandyopadhyay J, Ray SS. Influencing the shape recovery and thermomechanical properties of 3DP PLA using smart textile and Boehmite Alumina and Thermochromic dye modifiers. *Macromol*. 2022;2(3):485–99.

Qi X, Xiu H, Wei Y, Zhou Y, Guo Y, Huang R, Bai H, Fu Q. Enhanced shape memory property of polylactide/thermoplastic poly (ether) urethane composites via carbon black self-networking induced co-continuous structure. *Composites Science and Technology*. 2017;139:8–16.

Raviv D, Zhao W, McKnelly C, Papadopoulou A, Kadambi A, Shi B, Hirsch S, Dikovsky D, Zyracki M, Olguin C, Raskar R. Active printed materials for complex self-evolving deformations. *Scientific Reports*. 2014;4(1):7422.

Senatov FS, Niaza KV, Zadorozhnyy MY, Maksimkin AV, Kaloshkin SD, Estrin YZ. Mechanical properties and shape memory effect of 3D-printed PLA-based porous scaffolds. *Journal of the Mechanical Behavior of Biomedical Materials*. 2016;57:139–48.

Sharma A, Singholi AK. Shape memory and mechanical characterization of polylactic acid wood composite fabricated by fused filament fabrication 4D printing technology. *Materialwissenschaft und Werkstofftechnik*. 2021;52(6):635–43.

Shrivastava, A., Chohan, J.S. & Kumar, R. On mechanical, morphological, and fracture properties of sustainable composite structure prepared by materials extrusion-based 3D printing. *Journal of Materials Engineering and Performance*. 2023. https://doi.org/10.1007/s11665-023-08593-y

Singh G, Singh S, Prakash C, Kumar R, Kumar R, Ramakrishna S. Characterization of three-dimensional printed thermal-stimulus polylactic acid-hydroxyapatite-based shape memory scaffolds. *Polymer Composites*. 2020;41(9):3871–91.

Singh M, Singh R, Kumar R, Kumar P, Preet P. On 3D-printed ZnO-reinforced PLA matrix composite: Tensile, thermal, morphological and shape memory characteristics. *Journal of Thermoplastic Composite Materials*. 2022;35(10):1510–31.

Sorimpuk NP, Choong WH, Chua BL. Thermoforming characteristics of PLA/TPU multi-material specimens fabricated with fused deposition modelling under different temperatures. *Polymers*. 2022;14(20):4304.

Vardhan H, Kumar R, Chohan JS. Investigation of tensile properties of sprayed aluminium based PLA composites fabricated by FDM technology. *Materials Today: Proceedings*. 2020;33:1599–604.

Zarek, M. et al., 4D printing shape memory polymers for dynamic jewellery and fashionwear. *Virtual and Physical Prototyping*. 2016;11(4):263–270. doi:10.1080/17452759.2016.1244085

Zeng C, Liu L, Bian W, Liu Y, Leng J. 4D printed electro-induced continuous carbon fiber reinforced shape memory polymer composites with excellent bending resistance. *Composites Part B: Engineering*. 2020;194:108034.

Zhang F, Wen N, Wang L, Bai Y, Leng J. Design of 4D printed shape-changing tracheal stent and remote controlling actuation. *International Journal of Smart and Nano Materials*. 2021;12(4):375–89.

9 Process Capability Analysis for Thermoplastic Composite for Manufacturing of Sustainable Structures

Vishal Thakur, Rupinder Singh, and Ranvijay Kumar
Chandigarh University, Mohali, India

9.1 INTRODUCTION

Additive manufacturing (AM) is well-known by another term: "3D printing," which is the technique of constructing a structure or part in layer addition form. Any complex shape that is challenging to the machine can be produced using CAD/CAM technology (Wong and Hernandez, 2012). The fabrication of 3D-printed objects in the desired shape uses a variety of AM techniques, consisting of fused deposition modeling (FDM), stereolithography (SLA), material jetting, powder bed fusion, sheet lamination, and binder jetting, among others. According to American Society for Testing and Materials (ASTM) standards, each AM technique has a different process and material appropriateness (Tan et al., 2020). The materials considered for the process of AM are single polymers, ceramics, metals, and a mixture of more than one material (Bourell et al., 2017). Thermoplastic composites were used in structural applications and evolved rapidly due to their high strength, stiffness, and other mechanical properties. Polymers or composites reinforced with fiber have become common engineering materials. The incorporation of fibers in thermoplastic polymers has enhanced the mechanical properties (Vaidya and Chawla, 2008). The previous study has revealed the fabrication of the hemp/PLA based fiber composites (Xu et al., 2019). The authors have reported that PLA was reinforced with CF for composite structure formation. The tensile and flexural strength has increased by 13.8% and 164%, respectively (Li et al.,

DOI: 10.1201/9781003309123-9

2016). For load-bearing applications, flax fiber was incorporated into PLA. The tensile strength and stiffness of PLA-flax fiber composites were observed to be 72 MPa and 13 GPa, respectively, which is advantageous for load-bearing applications (Nassiopoulos and Njuguna, 2015). The Cordenka rayon and flax fibers were reinforced in PLA for investigation of mechanical properties. The tensile strength and impact strength found for PLA-Cordenka rayon fiber is 58 MPa and 72 kJ/m^2. For PLA-flax fiber composite, the highest Young's modulus observed is 6.31 GPa. The PLA-Cordenka fiber is a promising composite material and can be used in the automotive and electronic industries (Bax and Müssig, 2008). PLA-kenaf composites can be used for the applications of biodegradable composite materials. The tensile and flexural strength of PLA with reinforced kenaf fiber was observed at 223 MPa and 254 MPa. Also, in the time of four weeks, the composite weight decreased by 38% (Ochi, 2008). Composites made of PLA and natural fibers exhibited mechanical and biodegradable qualities that were comparable to those of synthetic composites. Whereas conventional fibers and synthetic polymers are exceedingly difficult to recycle and have very little sustainability, PLA-natural fiber composites are both easily recyclable and environmentally friendly (Siakeng et al., 2019). The PLA composites can be used for applications such as tissue engineering and drug delivery, automotive parts, aerospace and heritage structures (Kumar et al., 2022; Liu et al., 2020; Jung et al., 2011; Sanivada et al., 2020).

The literature is evident that very few studies were reported related to PLA-CF composite structure. This chapter aims to investigate the process capability analysis of PLA-CF composite for the development of sustainable structures. This study is an extension of the study reported by Thakur et al. (2023). In this study, an extension has been made for evaluating the process capability parameters of the PLA-CF-based sandwiched composite structures for industrial prospectives.

9.2 MATERIAL AND METHODS

The manufacturing of PLA-CF composites used CF and PLA filament (diameter: 1.75 ± 0.05 mm). Composites made of PLA-CF were manufactured by using the FDM method and CF was sandwiched in the middle of the PLA layers.

9.3 MANUFACTURING OF COMPOSITE STRUCTURES BY FDM PROCESS

The composites were manufactured by using PLA as matrix material with sandwiched CF by using the FDM process. Table 9.1 depicts that the PLA-CF composites were printed as per the design of the experiment (DOE) formed by using Taguchi L18 orthogonal array. The parameter setting used for manufacturing composite samples is 0.12 mm thickness of layer thickness, an infill pattern is linear, and 80% infill density. The set of the process parameters and method was adopted from the previous study reported by Thakur et al. (2023). The slicing software

TABLE 9.1
DOE for the Development of PLA-CF Composites

Sample No.	Type of Processing	Fiber Orientation	Nozzle Temperature (°C)	Bed Temperature (°C)
1	Pre-heat processing	0°	200	55
2	Pre-heat processing	0°	205	60
3	Pre-heat processing	0°	210	65
4	Pre-heat processing	45°	200	60
5	Pre-heat processing	45°	205	65
6	Pre-heat processing	45°	210	55
7	Pre-heat processing	90°	200	65
8	Pre-heat processing	90°	205	55
9	Pre-heat processing	90°	210	60
10	Post-heat processing	0°	200	65
11	Post-heat processing	0°	205	55
12	Post-heat processing	0°	210	60
13	Post-heat processing	45°	200	60
14	Post-heat processing	45°	205	65
15	Post-heat processing	45°	210	55
16	Post-heat processing	90°	200	65
17	Post-heat processing	90°	205	55
18	Post-heat processing	90°	210	60

Ultimaker Cura software package was used to slice all of these samples (Version 4.5). The parameters of the AM process for manufacturing PLA-CF composites include bed temperature, nozzle temperature, fiber orientation, and type of processing (pre-heat and post-heat processing).

9.4 TENSILE TESTING

Using PLA reinforced with CF at various settings, the tensile specimen samples were made. A 3D printer (Creality 3D Ender Pro) was used to make a total of 18 samples. As per ASTM D638 type IV specifications, the dimensions of manufactured PLA-CF composite tensile specimens were determined. The PLA-CF specimens were subjected to tensile testing utilizing the universal testing machine (UTM) (make: Shanta Engineering, maximum capacity: 5000 N) at a strain rate of 20 mm/min.

For the PLA-CF composite tensile specimens, the strength at peak and break were observed. For sample 2 (pre-heat processing, 0° fiber orientation, 205° nozzle temperature, and 60° bed temperature), the maximum strength at peak and break was 69.94 MPa and 62.95 MPa, respectively. The enhancement of mechanical characteristics is significantly affected by the fiber orientation. At 0° orientation, as the fibers are placed in a transverse direction or the same tensile direction, the tensile elongation of the composite is greater, which improves or enhances the strength of the composite.

9.5 DIMENSIONAL MEASUREMENTS

The setting parameters of the orientation of fiber is 0°, nozzle temperature at 205°C, and bed temperature at 60°C possessed better mechanical properties, so, ten samples were manufactured/repeated at this parameter setting. The process capability analysis was performed for these ten samples. The tensile specimen includes a grip section, gauge length section and thickness as shown in Table 9.2. So, the process capability analysis was done for these three sections of PLA-CF composite tensile specimens.

9.5.1 Process Capability Analysis for the Grip Section

The process capability analysis was done by using the process capability wizard software package. Figure 9.1(a) and (b) show the histogram and normal probability plot by using the dimensional values of the width of the grip section of ten samples (shown in Table 9.2). The Cp and Cpu values observed are 1.640 and 1.606, respectively (which are >1), for the grip section, as shown in Table 9.3. As the values of Cp and Cpu are >1, these results suggest that the dimensional accuracy for these composite samples is good.

9.5.2 Process Capability Analysis for Gauge Section

The histogram and normal probability for the gauge length section of composite samples are shown in Figure 9.2(a) and (b). The Cp and Cpu values that were observed are 2.072 and 1.029, respectively, for the width of the gauge length section of the composite sample. These values of Cp and Cpu lie above 1, which also shows that the dimensional accuracy of a section of the gauge length of composite samples is implemented (Table 9.4).

TABLE 9.2
Specific Dimensions of Tensile Specimens

Sample No.	Width of Grip Section (in mm)	Width of Gauge Length Section (in mm)	Thickness (in mm)
1	19.27	6.21	3.82
2	18.94	6.18	3.80
3	18.88	6.11	3.88
4	18.97	6.15	3.93
5	19.41	6.22	3.88
6	19.21	6.19	3.91
7	18.86	6.08	3.81
8	19.07	6.12	3.90
9	18.80	6.09	3.96
10	18.79	6.16	3.89

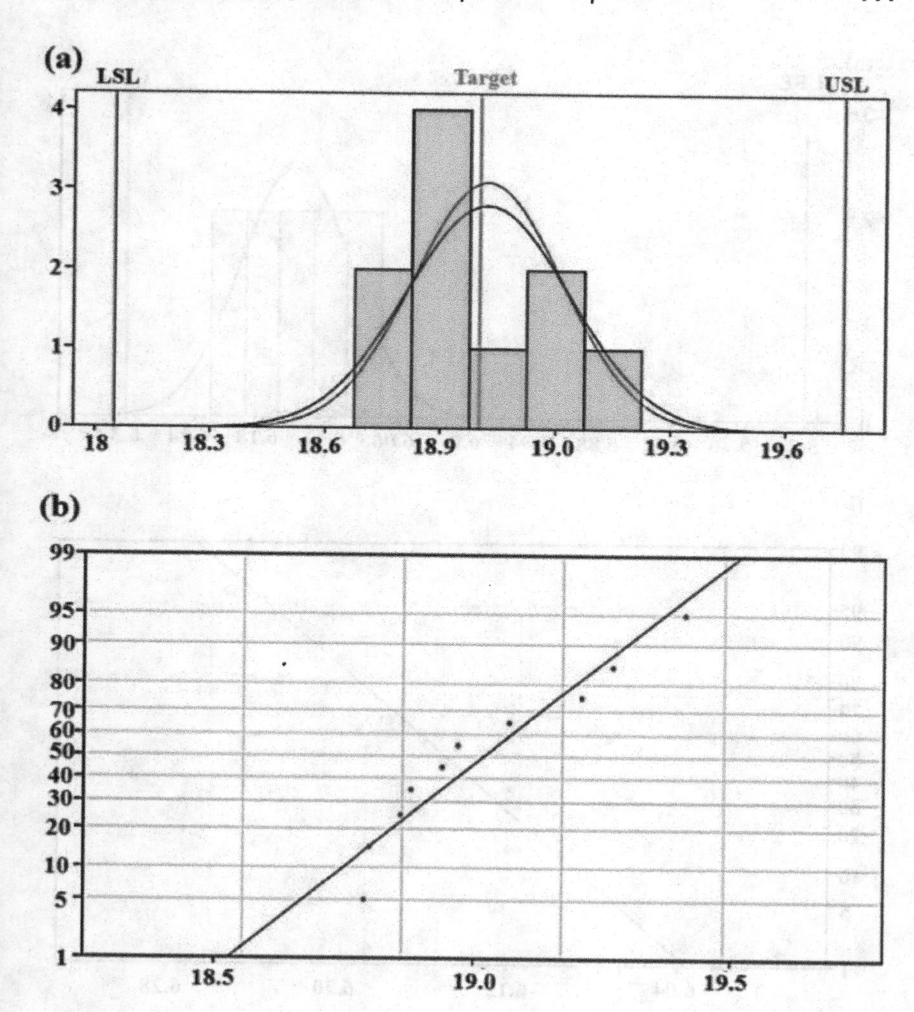

FIGURE 9.1 (a) Histogram (b) normal probability plot for width of grip section of PLA-CF tensile specimen.

TABLE 9.3
Process Capability Statistics for Width of Grip Section of PLA-CF Tensile Specimen

LSL	18.05
USL	19.95
AD Test	Passed
Cp	1.640
Cpu	1.606

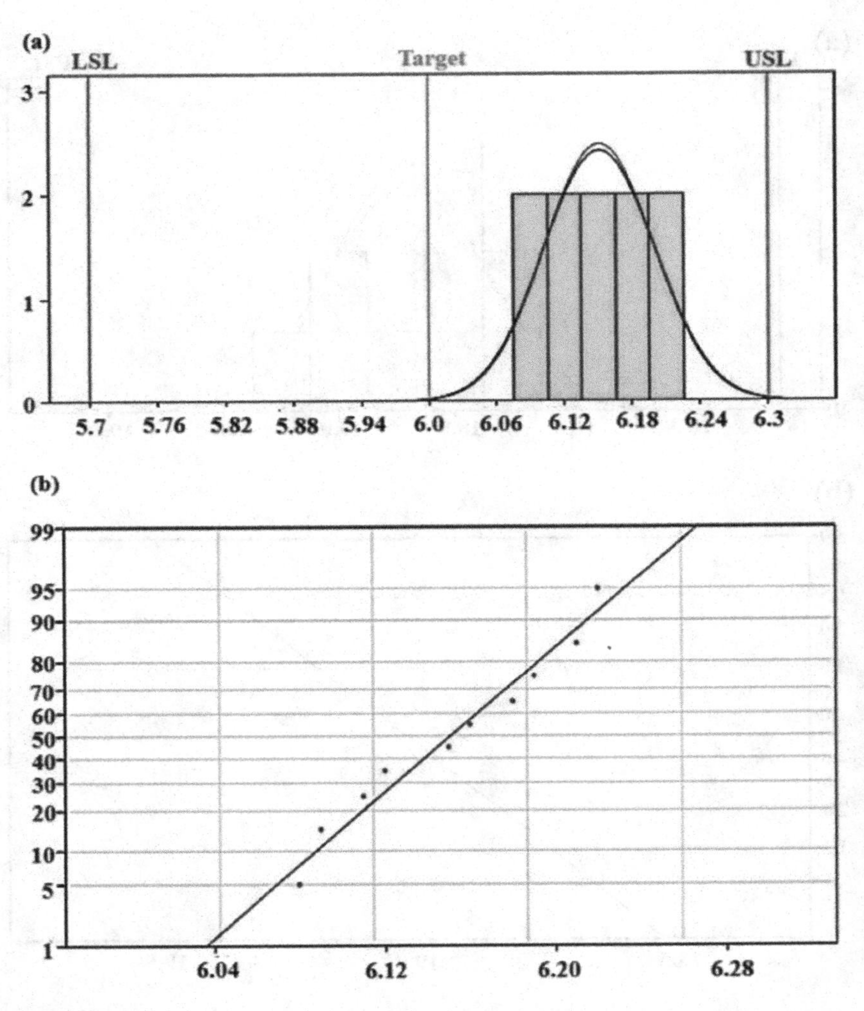

FIGURE 9.2 (a) Histogram (b) Normal probability plot for the width of the section of gauge length of PLA-CF tensile specimen.

TABLE 9.4
Process Capability Statistics for the Width of the Section of Gauge Length of PLA-CF Tensile Specimen

LSL	5.7
USL	6.3
AD Test	Passed
Cp	2.072
Cpu	1.029

9.5.3 Process Capability Analysis for Thickness

The histogram graph and normal probability plot for the thickness section of composite tensile specimens are shown in Figure 9.3(a) and (b). the Cp and Cpu values that were observed are 1.231 and 1.981, respectively (which are >1), for the thickness dimension of composite samples. As the values of Cp and Cpu are >1, the dimensional accuracy of the thickness section of composite samples is concerned (Table 9.5).

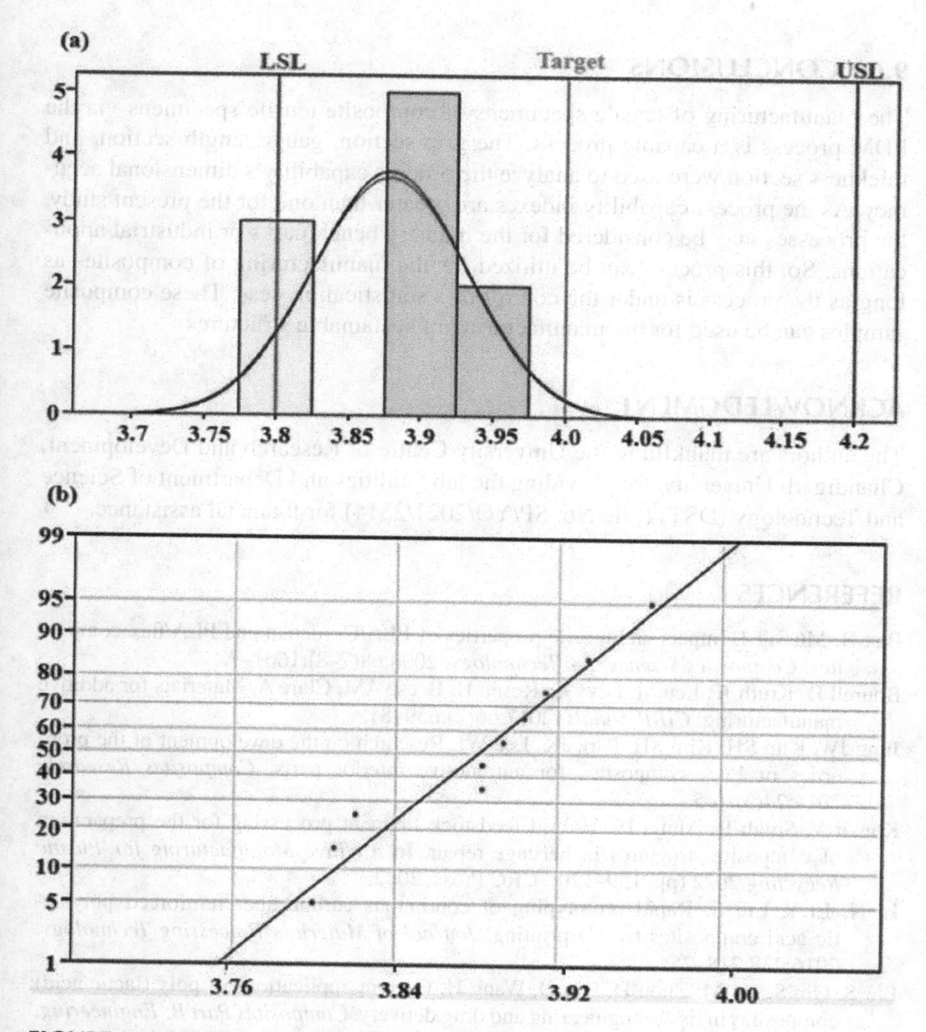

FIGURE 9.3 (a) Histogram (b) Normal probability plot for thickness section of PLA-CF tensile specimen.

TABLE 9.5
Process Capability Statistics for Thickness Section of PLA-CF Tensile Specimen

LSL	3.8
USL	4.2
AD Test	Passed
Cp	1.231
Cpu	1.981

9.6 CONCLUSIONS

The manufacturing of tensile specimens of composite tensile specimens via the FDM process is a capable process. The grip section, gauge length section, and thickness section were used to analyze the process capability's dimensional accuracy. As the process capability indexes are greater than one for the present study, the processes may be considered for the industry benchmarks or industrial applications. So, this process can be utilized for the manufacturing of composites as long as the process is under the control of a statistical process. These composite samples can be used for the manufacturing of sustainable structures.

ACKNOWLEDGMENT

The authors are thankful to the University Centre of Research and Development, Chandigarh University, for providing the lab facilities and Department of Science and Technology (DST) (File No. SP/YO/2021/2514) for financial assistance.

REFERENCES

Bax B, Müssig J. Impact and tensile properties of PLA/Cordenka and PLA/flax composites. *Composites Science and Technology*. 2008;68(7–8):1601–7.

Bourell D, Kruth JP, Leu M, Levy G, Rosen D, Beese AM, Clare A. Materials for additive manufacturing. *CIRP Annals*. 2017;66(2):659–81.

Jung JW, Kim SH, Kim SH, Park JK, Lee WI. Research on the development of the properties of PLA composites for automotive interior parts. *Composites Research*. 2011;24(3):1–5.

Kumar V, Singh R, Ahuja IS. Hybrid feedstock filament processing for the preparation of composite structures in heritage repair. In *Additive Manufacturing for Plastic Recycling 2022* (pp. 159–170). CRC Press, 2022.

Li N, Li Y, Liu S. Rapid prototyping of continuous carbon-fiber reinforced polylactic acid composites by 3D printing. *Journal of Materials Processing Technology*. 2016;238:218–25.

Liu S, Qin S, He M, Zhou D, Qin Q, Wang H. Current applications of poly (lactic acid) composites in tissue engineering and drug delivery. *Composites Part B: Engineering*. 2020;199:108238.

Nassiopoulos E, Njuguna J. Thermo-mechanical performance of poly (lactic acid)/flax fibre-reinforced biocomposites. *Materials & Design*. 2015;66:473–85.

Ochi S. Mechanical properties of kenaf fibers and kenaf/PLA composites. *Mechanics of Materials*. 2008;40(4-5):446–52.

Sanivada UK, Mármol G, Brito FP, Fangueiro R. PLA composites reinforced with flax and jute fibers—A review of recent trends, processing parameters and mechanical properties. *Polymers*. 2020;12(10):2373.

Siakeng R, Jawaid M, Ariffin H, Sapuan SM, Asim M, Saba N. Natural fiber reinforced polylactic acid composites: A review. *Polymer Composites*. 2019;40(2):446–63.

Tan LJ, Zhu W, Zhou K. Recent progress on polymer materials for additive manufacturing. *Advanced Functional Materials*. 2020;30(43):2003062.

Thakur V, Kumar R, Kumar R, Singh R, Kumar V. Hybrid additive manufacturing of highly sustainable Polylactic acid-Carbon Fiber-Polylactic acid sandwiched composite structures: Optimization and machine learning. *Journal of Thermoplastic Composite Materials*. 2023:08927057231180186.

Vaidya UK, Chawla KK. Processing of fibre-reinforced thermoplastic composites. *International Materials Reviews*. 2008;53(4):185–218.

Wong KV, Hernandez A. A review of additive manufacturing. *International Scholarly Research Notices*. 2012;2012.

Xu Z, Yang L, Ni Q, Ruan F, Wang H. Fabrication of high-performance green hemp/polylactic acid fibre composites. *Journal of Engineered Fibers and Fabrics*. 2019;14:1558925019834497.

10 Sustainable Production of Sponge Iron through Direct Reduction Process

Waste Recovery from Mill Scale

A. A. Adeleke
Nile University of Nigeria, Abuja, Nigeria

P. P. Ikubanni
Landmark University, Omu-Aran, Nigeria

J. K. Odusote and H. O. Muraina
University of Ilorin, Ilorin, Nigeria

Harmanpreet Singh
Indian Institute of Technology Ropar, Rupnagar, India

D. Paswan and M. Malathi
National Metallurgical Laboratory, Jamshedpur, India

10.1 INTRODUCTION

There has been an increase in the yearly estimated per capita consumption of iron and steel from 5 kg in 1968 to 130 kg in 2012 in Nigeria (Ohimain, 2013). In 1958, the Nigerian steel sector took off, and to date (over 60 years), the sector is still struggling for stability despite a huge investment of over $7 billion (Ohimain, 2013). The iron and steel sector is economically viable though with various technological and financial challenges emanating from iron ore beneficiation to liquid iron phases. These challenges are summed up by a lack of quality

DOI: 10.1201/9781003309123-10

leadership in Nigeria. Proper harness of the country's mineral resources is the focus of this review. Nigeria has over three billion tons of coal and iron ore, 700 million tons of limestone and 187 billion SCF of natural gas, which are significant pointers that direct reduced iron (DRI) and can solve the country's dangling challenges in producing pig iron from blast furnace (BF) (Agbu, 2007; Ohimain, 2013; Wiedmann et al., 2015; Zachariah et al., 2022). The need for iron and steel is projected to increase exponentially in the coming years. This assertion is based on the tremendous graduation from 2015 to 2016, as shown in Figure 10.1(a–b) (Agbu, 2007; Kumar, 2014; Fick et al., 2014; Mitra Debnath and Sebastian, 2014; Hashem et al., 2015; Zachariah et al., 2022; WCST, 2022). In the production of iron and steel, there are two unique routes: BF extraction to basic oxygen furnace (BOF) refining and direct reduction (DR) to electric arc furnace (EAF) (Fick et al., 2014). BF is a steel plant that involves sintering sections (pelletization), coke ovens, BF, and BOF (Fick et al., 2014). Meanwhile, BF is enshrined with lots of financial burden and charges (raw materials) are of high grade. Coking coal is necessary to support the burden of the BF (Mitra Debnath and Sebastian, 2014; Odusote et al., 2019). The steel per ton produced by BF is low and likely to continue to decline relative to the ton of steel produced via EAF, for which DRI is the raw material (Goldring, 2003; Olayebi, 2014; Mitra Debnath and Sebastian, 2014; Odusote et al., 2019; WCST, 2022). DRI → EAF is an alternative route through which iron and steel can be produced (Kirschen et al., 2011). This channel is considered an alternative to BF due to its uniqueness, nature of charges, and less financial burden. DRI is made possible by a reducing an agent known as a reductant. Reductant serves as a source of heat and supplies carbon for the initiation of Bouduard's reaction, which is important in the reduction of the charged iron ore or any other iron-bearing materials (Kirschen et al., 2011; Ahmed, 2018; Ramakgala and Danha, 2019). Reductant sources could be via natural gas, primarily the mixture of H_2 and CO, or non-coking coal. Iron ore or iron-bearing materials (such as mill scale) are reduced in solid state around 800°C–1050°C (1073–1323 K) by non-coking coal or natural gas in the direct reduction process. Thus, direct reduction has been characterized as energy efficient (Grobler and Minnitt, 1999; Dutta and Sah, 2016).

Moreover, DRI is not only an alternative for steel scrap in electric arc furnaces (EAF/IF) but also an important charge for good-quality steel production (Grobler and Minnitt, 1999; Dutta and Sah, 2016; Mishra and Roy, 2016). EAFs presently account for about 28%–30% of the steel produced in the world. The quest for qualitative steel at a realistic cost necessitates the exploration of a substitute for steel scraps. However, with the emergence of DR, the problem has become more glaring. As an addendum, the start-up capital per running cost investment of direct reduction plants is lower compared to combined plants (BF → BOF) and direct reduction plants because these features are more suitable and viable for developing countries. The product of direct reduction is referred to as sponge iron. It is called sponge iron because it is solid pellets/lumps with voids filled with air. This chapter reviews various iron-bearing materials for DRI (sponge iron) vis-à-vis their production routes, properties and, importantly, the future trend of sponge iron and applications (Dutta and Sah, 2016).

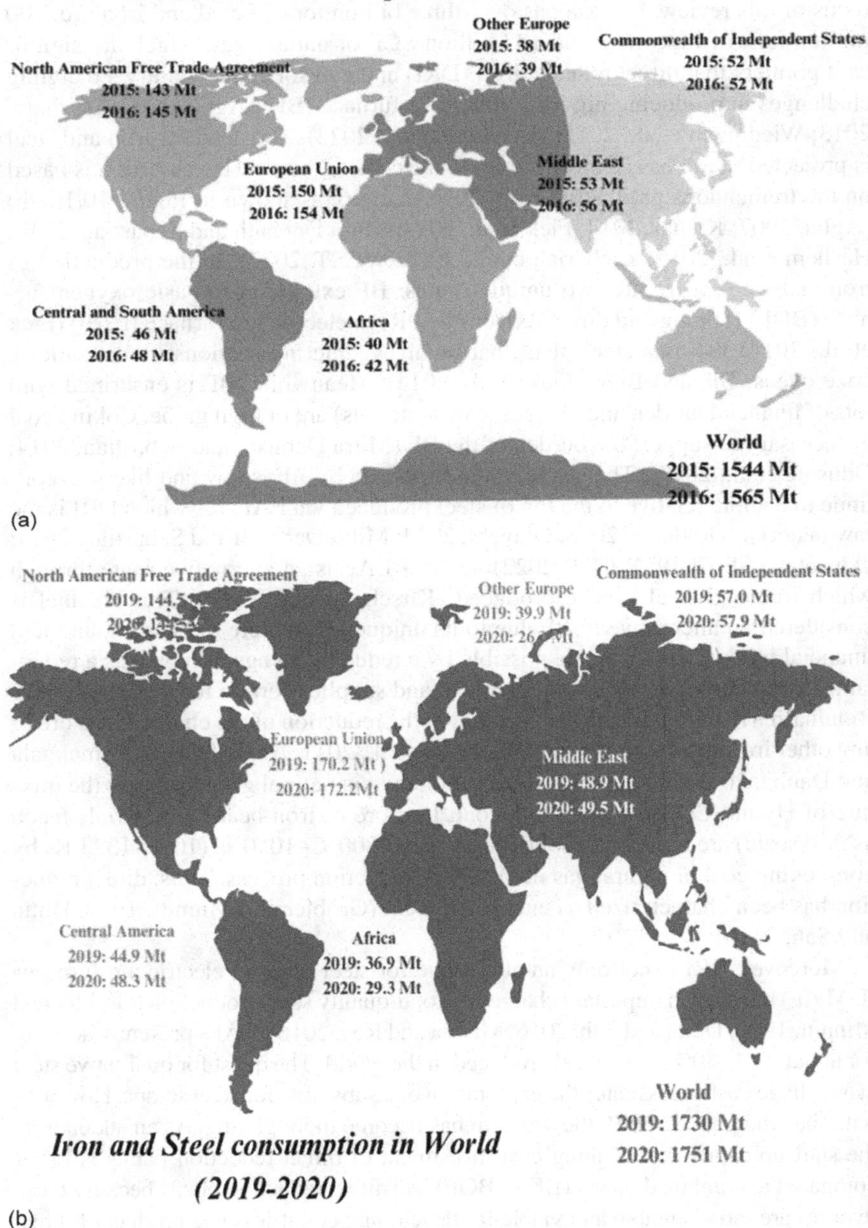

FIGURE 10.1 (a) Iron and steel consumption for the world between 2015 and 2016. (b) Iron and steel consumption for the world between 2019 and 2020.

Source: Out of courtesy (author's work).

10.2 DIRECT REDUCED IRON (DRI)

DRI is a highly porous iron that is produced via a direct reduced (DR) process. DR-processed iron oxides could either be solid-to-solid or solid-to-gas reactions. It is a process in which oxygen is driven out from the iron oxide using natural gas or coal as a reductant below the melting temperature of the ore (lump or fine). The process has nothing to do with the outer shape of the pellets/lumps since it is carried out at comparably lower temperature points (Grobler and Minnitt, 1999; Kumar et al., 2008; Hashem et al., 2015; Mishra and Roy, 2016; Ramakgala and Danha, 2019). However, there is the possibility of the feedstock to greatly reduced (weight) by about 27%–30% due to the removal of excess oxygen. In addition, the color of the charges after the DR process changes from red color to black color (Figure 10.2) while density falls between 3.50×10^3 and 4.40×10^3 kg/m^3. Originally, the density of pure iron was around 7.8×10^3 kg/m^3. Thus, the process accounts for about a 45%–56% reduction in true volume. The changes in volume are apparent in the sponge iron pieces produced, and this is evident in the pores produced throughout the interior. If sponge pellets or iron becomes wet, it can overheat significantly and emit hydrogen gas. Thus, the process must be carried out under inert conditions (Hashem et al., 2015; Mishra and Roy, 2016; Ramakgala and Danha, 2019).

FIGURE 10.2 DRI.

Source: Ahmed, H. (2018). New trends in the application of carbon-bearing materials in blast furnace ironmaking. *Minerals,* *8*(12), 1–20. https://doi.org/10.3390/min8120561.

10.2.1 Solid-Solid Reduction Process

The solid-solid reduction process is one of the ways through which DRI can be recovered. Coal-based DR process possesses some advantages over the gas-based DR process. In most cases, the process does not require coking or quality coal for it to take place. DR processes use non-coking or lean-graded coal. The plants can be mounted at a lower capacity and possibly at places where coal and iron ore reserves or deposits are low. The process can be operated at a small scale and this makes them suitable for developing countries. Despite all the beneficial attainment of this process, the lower economy scale, lower productivity (between 0.5 and 0.9 t/m^3/day), and low carbon content (< 1.0%) in the product marred the process. To have an optimum operation, energy efficiency must be between 16.0 and 21.0 GJ/t, which is visibly high. Sponge iron cannot be fed into a steelmaking furnace directly due to the presence of some ash and char residues (White, 1929; Umadevi et al., 2009; Adebimpe and Akande, 2011; Anand et al., 2016).

10.2.2 Solid-Gas Reduction Process

Gas-based process of direct reduction is obviously cheaper, especially when considering the capital cost/ton of capacity of the plant. The cost/capital ton of capacity for coal-based plants is 1.8–2 times of plants operated on gas. Besides, the production rates of coal-based plants are extremely lower compared to gas-based direct reduction plants. For gas-based plants, the production rate can be 11 t/m^3/day compared with 0.5–0.9 t/m^3/day for coal-based plants. Hence, there is high productivity. In addition, the quality of DRI produced based on the degree of metallization and carbon content is better for gas-based plants. Moreover, the degree of energy consumed during the process is quite efficient, and there is better plant availability. Gas-based plants are substantially standardized, coupled with gas being a clean source of fuel, thus, the challenges with huge ash formation are eradicated, which makes producing quality iron hugely possible. Maintenance problems of the plants are obviously limited in this process, unlike the coal-based process. The issue of environmental pollution is inherently ameliorated with gas being a clean source of fuel residues (White, 1929; Umadevi et al., 2009; Adebimpe and Akande, 2011; Anand et al., 2016).

10.3 IRON-BEARING MATERIALS

10.3.1 Iron Ore

Ores are naturally occurring deposits in the earth's crust from which engineering materials are economically recovered or extracted. Ores of iron are generally referred to as oxides or iron ore. Iron ore is a major ingredient to produce iron and steel. Iron ores are mineral rocks from which extraction of metallic iron is possible. Iron ores are rich in oxides and likewise differ in their visual nature. Some ores are dark gray in color, while some are brightly yellowish, intense purple to rusty red. They can be in form of magnetite (Fe_3O_4) – 72.4% Fe, hematite (Fe_2O_3) – 69.9% Fe, goethite (FeO (OH)) – 62.9% Fe, limonite ($FeO(OH){\cdot}n(H_2O)$) – 55% Fe or siderite ($FeCO_3$) – 48.2% Fe (White, 1929; Ramakgala and Danha, 2019; Odusote et al., 2019). Hematite or magnetite are predominantly banded with 60% iron and are referred to as direct shipping or natural ore. The implication is that hematite or magnetite can be fed directly into BFs for iron production since iron ores must contain around 60 wt.% Fe to be fit as feedstock for BF operations. It has been reported that feedstocks must have over 67 wt.% Fe for some new processes for iron recovery (Goldring, 2003; Baba et al., 2006; Agbu, 2007; Adebimpe and Akande, 2011). Beneficiation of ore will be followed by smelting to make pig iron and subsequently processed to obtain steel (Baba et al., 2006; Agbu, 2007; Adebimpe and Akande, 2011). Iron ore is a vital part of the world economy and, as such, will compete heavily with oil in the future. Pig iron is the product of BF, from which iron ore is an active input. Nigeria, as a country, is blessed with huge iron ore and other mineral resources deposits such as limestone, salt, clay, gypsum, barites, and raw materials for energy like coal, oil, gas, and bitumen. Nigeria is rich in iron ore, with its richness displayed in more than 3 billion metric tons of its deposits across the country (Bamalli et al., 2011; Ohimain, 2013; Dutta and Sah, 2016). The iron ore reserves in Nigeria, their locations, chemical constituents, and percentages of Fe and hematite are shown in Table 10.1.

10.3.2 Iron Ore Fines (Slime)

Making recourse to environmental impact during the process of ore mining vis-à-vis beneficiation from the mining industries/sites is important. The huge quantity of waste or reject recovered during the treatment of iron ore is becoming conspicuous. In India, for example, about 10–29 million metric tons of this waste is generated yearly, which has contributed to the environmental challenges of the country. These rejects are referred to as ore fines or slime (Pal et al., 2010; Roy and Das, 2013; Filippov et al., 2014; Jena et al., 2015; Sarkar et al., 2017; Roy et al., 2020). They are usually regarded as lower-grade ore. The percentage of iron content in slime is between 48 and 56%, which is relatively high compared

TABLE 10.1
Iron Ore Locations, Compositions, and Reserves across Nigeria (Bamalli et al., 2011)

Locations	Reserved	Compositions (%)								
	(Tons)	Fe	Fe_2O_3	SiO_2	Al_2O_3	CaO	MgO	P_2O_5	MnO	TiO_2
Agbaja	2×10^9	45–54	62.64	8.55	9.60	0.72	0.38	4.16	0.14	0.37
Chokochoko	7×10^7	37.43	29.41	0.62				0.05	0.52	
Nsude Hills	6×10^7	37.43								
Itakpe	$(2–3) \times 10^8$	38–45	53.10	44.80	1.00	0.30	0.20	0.05	0.05	0.10
Agbade Okudu	7×10^7	37.43								
Ajabanoko	3×10^7	35.61	47.74	0.41				0.11	0.05	0.06

to some iron ores. Al_2O_3 (6.78%) and SiO_2 (5.8%), which are impurities in slime, make it unsuitable for application in the BF. However, slime in its natural state may be immaterial not until when it is properly beneficiated. Two alternative chemical treatment methods have been recommended by Pal et al. (2010) to place value on iron ore fines. These include acid extraction of iron, which is to be followed by alkali treatment, and alkali extraction of Al_2O_3 and SiO_2 from iron ore slime. It was reported that alkali extraction improved iron content to 65% w/w, while acid extraction could only improve to 63% w/w. Electro-winning of iron from oxalic acid solution using a steel cathode and a lead anode was surrounded by a synthetic cloth diaphragm in the experiment. High-purity iron was deposited on the cathode surface based on the experimental setup. It was concluded that the elemental iron produced by electro-winning can be directly used for steelmaking (Pal et al., 2010). Iron ore slime with 57% Fe can be beneficiated using hydrocyclone to a product with 64 Fe and a recoverable 49% iron (Das et al., 1992). Moreover, column flotation can be used to upgrade the iron content of ore slime from 57.5% to 65.5% to have a recovery of 50% Fe (Das et al., 2005). Kumar and Mandre (2017) used a selective flocculation approach to recover iron from iron ore slimes. Initial chemical and mineralogical characterization were carried out on the samples. Haematite and goethite were found to be the principal iron-bearing minerals, while quartz and kaolinite were the gangue minerals. It was reported that Fe grade can be improved from 58.24 to 64.60% Fe at a recoverable iron of 66.33% with the use of polyacrylamide as flocculants. Guar gum flocculants gave a Fe grade of 63.20% at a recoverable iron of 68.04%. Iron ore slime can be useful as an adsorbent to remove Pb (II) and Hg (II). It can also be used in the basic oxygen steelmaking process as a coolant.

10.3.3 Slag

The production of iron or steel is linked with the generation of solid waste material known as slag. In practice, the production of a metric ton of steel leads to the generation of 1 ton of solid waste. Big steelmaking plants produce around 29 million metric tons of solid waste materials. Predominantly, slag is mainly used as aggregate; however, it has now been used for cement production and fertilizer feedstocks (Ogbesode et al., 2023). Slag is the siliceous melt, which is obtained in considerable amounts from the production of iron and steel. These are the solid materials formed on the surface of melt when it cools. Production of ferrochrome leads to substantial tonnages of slag. The smelting of non-ferrous metals is associated with slag production. The slag produced from such melting are different and needs to be studied separately and cautiously (Chand et al., 2015; Wang, 2016; Kumar and Mandre, 2017). Clinker, ash and even colliery wastes are referred to as slag. Slag is produced from impurities (gangue) that were initially bound to iron ore. However, nowadays, there is a large contribution from various materials input aimed at removing some specific detrimental element (Chand et al., 2015; Wang, 2016; Kumar and Mandre, 2017). Lime, for example, is an important material for BF operation. It enhances the fluidity of the slags, militates the loss of iron as oxides within the slag, and negates the action of coke by removing the excess sulfur in the reaction instituted by coke. Further desulphurization of steel may also generate a small proportion of slag in the other plants for further steel processing. In some climes, slag is used for sand filling in low areas. The chemical properties of slag are comparable to natural aggregates used in civil engineering construction works. Thus, slag from steel can replace these aggregates, as they are considered wastes, which are far cheaper. Slag from steel are useful aggregates for filters in wastewater treatments; they can be used to remove heavy metals from wastewater (Grobler and Minnitt, 1999; Chand et al., 2015; Wang, 2016; Kumar and Mandre, 2017).

10.3.3.1 Blast Furnace Slag

The composition of any iron oxide is naturally composed with several percent of other constituents, such as silica and alumina, whose excess results in slag formation in the BF burden. For alumina, the level or proportion usually varies according to the sources of such ore. Another source of slag formation is the ash from coke used in the reduction of iron ores in the BF, as well as additives such as lime inclusion. The inclusion of lime enhances the metallurgical process, while the addition of titanium (another additive) is periodically used to militate the effect of wear in the lining of the furnace. Slags in some climes are regarded as waste and traditionally disposed of as such, probably as landfills. In addition, their effect on the environment in terms of pollution is alarming. Therefore, it is important to place value on them via the recovery of metals (Qin et al., 2012). BF slag, when it is properly air-cooled, is rock-like in nature, though a bit more porous than most natural rocks. This feature makes BF slag meritorious to be used as aggregate in sewage treatment beds. Though with the desired chemical composition in mind, pulverized BF slag can be added to Portland cement for the production of concrete that is satisfactorily inexpensive and technically feasible, bearing in mind the percentage alumina, which, as earlier stated, is highly detrimental when it is obviously high in the iron ore composition. BF slag is said to contain large amounts of sensible heat (Qin et al., 2012). The temperature of BF slag is 1,773K, and its enthalpy is around 1,700 MJ/ton. Granulating molten BF slag by rotary multi-nozzle cup atomizer and pyrolyzing printed circuited board has been proposed for heat recovery from hot BF slag. Dry granulation and pyrolyzation were used to verify the feasibility of the waste heat-recovery method. A pyrolysis process was used to convert the energy of hot BF slag into chemical energy. An enormous quantity of combustible gas, such as CH_4, CO, C_mH_n, and H_2, was generated during the process (Qin et al., 2012).

10.3.3.2 Steelmaking Slags

Steel slag is generated when oxygen is blown into the molten metal in a BOF. Open-hearth and Bessemer processes were the common ones before EAF and BOF. Compared to the Bessemer process, where air is blown through the molten metal, heated air and fuel are blown across the surface of molten metal that is being held in a vessel for the open-hearth process. They have a total output of about 10%–15% by weight of the steel (Huiting and Forssberg, 2003; Wang, 2016). However, with the emergence of the latest technologies, the production cost is one big reason for discarding these old approaches. It is important for steel to have clarity in quality in terms of production rather than quantity. Steels abound with impurities, and lots of undesirable elements display failure in service catastrophically and, in the long run, make producers economically run at a loss. To solve some of these impurities challenges in steels, fluxing materials such as lime may be added as part of the active burden in the furnace to accelerate the slag formation. Commonly among the class of fluxes are calcium fluoride and certain borates. In some applications, red mud from the processing of alumina has also been used. Whichever way, any flux intending to be used must be environmentally friendly (White, 1929). In steel slag, the main elemental chemical compositions are SiO_2, CaO, Fe, MnO, and MgO (Marquardt et al., 2014). Normally, Fe amounted to around 7%–10% in the overall composition of steel slag, as well as in iron oxide and iron-bearing minerals. The surest and most technical way of separating iron from slag is via mineral processing and possibly by recycling them as feedstock for steelmaking, BF, and sintering. To reduce the financial burden associated with the production of steel, active ingredients such as limestone, dolomite, and manganese ore may be substituted for MnO, CaO, and MgO in steel slags, especially when they are significantly high, whereas P_2O_5 and S are present in large amounts in some steel slags iron concentrate. This is a disadvantage to recycling such steel slag for iron and steel recovery (Huiting and Forssberg, 2003).

10.3.3.3 Basic Oxygen Steelmaking Slag

Pig iron (hot metal or molten iron) from BF is but one of the long processes to achieving quality steel. Recovering quality steel will be determined based on the extent of reducibility of excess elements such as carbon, silicon, and phosphorus, and this can be achieved through processes such as LD (Linz-Donawitz) and basic oxygen processes (Marquardt et al., 2014; Wang, 2016; Kumar and Mandre, 2017). Magnesia is needed as the refractory lining in the basic processes to resist the lime-rich slag. The process of mitigating the excess elements present in the melt is a function of heat. The hot oxygen is introduced from the top of the furnace into the steel, and latent elements are oxidized in an anticipated manner. Carbon breaks out as carbon monoxide, while silicon becomes SiO_2 and finds its way into the slag. Phosphorus, on the other hand, is normally present in the melt only but a small amount of around 0.1%–0.2%, and this still calls for close monitoring. Phosphorus is removed by the addition of a designated amount of lime to each heat in the furnace. The removal of phosphorus is initiated after the dissolution of the coarse lime. There is always an excessive dust loss for finer lime. In addition, the lower lime/silica ratio in slag gives reduced phosphorus removal, though the higher ratio is not fluidized enough. Fluorspar (calcium fluoride) can be added to accelerate slag formation, though it will increase the cost of production (Marquardt et al., 2014; Wang, 2016; Kumar and Mandre, 2017). It was earlier stated that slag formation is a function of heat. Heat accounts for the lime/silica ratio, which is expected to be around 3% or 4%. The grade of steel being produced is also largely dependent on heat and may be guided mainly by the desired amount of phosphorus (Qin et al., 2012). Large amounts of free line may remain unreacted, which can react with water and disrupt the aggregate or structures that it has been used for in real applications. This is a major concern. There have been numerous studies on the ways to mitigate or stop the reactive nature of these slags in the application world. There is an evolution in several areas of steelmaking processes, which makes optimization important for a proper understanding of how to modify slag properties when obtained from steelmaking vessels (Sinha et al., 2020; Pal, 2019; Bechara et al., 2018). So, the introduction of siliceous additives can fast-track the reaction of the melt with free lime, but the particles must not be surrounded by a chilled coat of solid slag immediately. This will prevent any continuous reaction within the melt. Weathering of slag prior to utilization can lead to hydration of free lime so that the possibility of expansion or any kind of reaction is decreased before such slag is used as an aggregate (Li et al., 2015).

10.3.4 Mill Scale

The oxide waste that is produced when casting, rolling, forging, or making steel is called mill scale. Mill scale contains a high quantity of iron (65%–70%). Recycling of large amounts of mill scale is an in-house consumption using an integrated steel plant. However, there is no commercial process available for its use in the secondary sector of the industries (Umadevi et al., 2009). This is the reason for its unhealthy dumping or exportation at low/cheap prices. Mill scale and steel production in any clime are expected to increase concurrently as the more steel produced, the more mill scale is generated (Anand et al., 2016; Ramakgala and Danha, 2019). India, between 2003 and 2004, produced a mill scale of about 0.7 MT, a figure considered to be high when compared to steel production (≥25 MT), as shown in Table 10.2. Fortunately, between 2009 and 2010, there was significant progress in steel production in the country (India) to about 60 MT. This time around, between the latter years, 1.20 MT of mill scale was reported to be generated. Table 10.2 shows the trend increment in the steel/mill scale proportion of India (Anand et al., 2016).

The recycling of steel plant mill scale through the iron ore pelletization process was studied by Umadevi et al. (2009). It was reported that the addition of mill scale (10%) gave the optimal physicochemical and metallurgical characteristics for the pellet. Researchers have also examined the recycling of steel plant mill scale through iron ore sintering plant. It was concluded increase in the addition of mill

TABLE 10.2
Trend Increment in Steel/Mill Scale Proportion of India (Anand et al., 2016)

Year	Steel Production	Mill Scale Generation
2003–2004	25 MT	0.7 MT
2009–2010	60 MT	1.20 MT
2013–2014	70 MT	1.45MT

scale to the sinter elevated the total amount of Fe and FeO contents within the sinter. The increase caused a reduction in sinter productivity because sinter bed permeability was inhibited by the addition of an increased mill scale. The increase in mill scales also caused a reduction in degradation index and reducibility because of the increase in FeO content within the sinter. The use of a 40–50 kg mill scale per ton can yield all the other desired properties for sinter safe its productivity. Anand et al. (2016) investigated the production of sponge iron from low-grade iron ore and mill scale through simulating tunnel kiln. Process development feasibility study for the production of direct reduced iron using waste/low-grade iron ore (slime), mill scale and Jhama coal was studied. The reduction kinetics of the pellets slime and Jhama coal were likewise obtained. Based on reduction kinetics, the process for pellet reduction with Jhama coal was optimized to produce highly metalized direct reduced iron. The kinetic theory shows that waste iron ore slime can thermally be reduced to quality iron as high as 92% metalized property that can be used in EAF.

10.4 REDUCTANTS (REDUCING AGENTS)

The reducibility of iron-bearing materials to quality sponge iron is technically made possible under the influence of heat around 450°C–1100°C. At this low temperature comparatively, the reductant is usually a combination of H_2 and CO rather than just H_2 (White, 1929; Adebimpe and Akande, 2011). It was equally reported that iron ore could be reduced at 800°C–1050°C in the solid state by either reducing gas (CO + H_2) or coal (White, 1929; Adebimpe and Akande, 2011; Bechara et al., 2018).

10.4.1 Reductant: Coal and Coke

The journey to revolutionize the country's iron and steel sector will rather remain a mirage without referring to metallurgical coke. Metallurgical coke is an indispensable feedstock for BF setup (operations). A quality coke should be able to give room to smooth descent of the BF burden with as little degradation as possible while providing high thermal energy, metal reduction, the best permeability gas flow and molten products, and the lowest number and quantity of impurities. Furthermore, the addition of quality coke into BF will give high productivity, reduce hot metal cost, and drastically reduce the rate of coke included in the overall production. Like a coin with two sides, both coke and non-coking coals are products of the same source. Coking coal is coal that, when heated in the absence of oxygen, leaves solid, coherent remains with have metallic gray sheen. They possess all the physicochemical properties in the coke when it is commercially produced. Coke is produced from such coking coal (Bin et al., 2013; Kumar, 2014). Whereas a non-coking coal leaves a solid coherent residue that may not possess the physicochemical properties of the coke. The non-coking coals are those coals that may form solid remains but may not be suitable to produce coke. This coal may form aggregate, which will not meet the physicochemical properties required of a good coke. In summary, natural gas and coal, which can be used as an energy input during the process of direct reduction of iron, have been classified as a non-renewable energy source (fossil fuel) and, thus, may be tending toward extinction, as indicated by researchers. Finally, to mitigate the effects of greenhouse gases, which occur mostly because of the combustion of fossil fuels, there is a need to tap into the bio-wastes (agro-wastes) that are abundantly available in Nigeria (Bin et al., 2013; Odesola et al., 2013; Kumar, 2014).

10.4.2 Reductant: Bioenergy/Biomass

The need to mitigate against climatic challenges makes the world pay attention to alternative sustainable energy. Biomass is a type of renewable source of energy generated from processed agricultural wastes or residues and capable of dispensing energy to full capacity as much as either coal or any form of conventional fuel can. It is also an important source of energy and probably the most abundant fuel worldwide after coal, oil, and natural gas (Odusote and Muraina, 2017). Biomass currently contributes the greatest amount of renewable energy the globe consumes, with the capacity to generate 7,000 MW (Odusote and Muraina, 2017; Ajimotokan et al., 2019; Adeleke et al., 2019a). By its natural diversity, biomass is useful as renewable fibers in paper industries and several other areas. The essence of biomass on a scale of usability can be effectively compared as a first order of magnitude. The energy equivalent of cereals in the world is 31.3 EJ, while all the commercial boles are 14.3 EJ. Fuel wood and charcoal are commonly used for domestic cooking and heating, with an estimated 15.3 EJ in the developing nations. The usable portion of the current biomass applications amounts to 60.9 EJ (2/3 of US energy consumption, 1 quad = 1.055 EJ) (Costello and Chum, 1998). Biomass likewise has a substantial potential to increase the production of electricity, heat, and fuels for transport and other uses. Biomass energy could provide an upgrade in energy security and trade balances, can substitute imported fossil fuels, and can notably lead to reductions in greenhouse gas emissions if properly handled. In sub-Saharan Africa (SSA), the economy is largely dependent on agriculture, with the exception of Nigeria, South Africa and a few others that massively depend on fossil fuel as a major source of economy. In this part of the world, the huge agricultural wastes generated yearly cannot be said to have been hugely explored. In most cases, they are either used as landfills or openly discarded without making recourse. The huge agro-waste generated yearly in SSA because of high agricultural investments can be backwardly integrated and substituted to produce an energy source that will be magnanimous enough to be used by this generation and those to come. Several studies have been done, and other studies have been carried out using agro-wastes in the generation of clean energy for domestic and industrial utilizations (Ikubanni et al., 2019; Adeleke et al., 2019b, 2021a, 2021b, 2021c, 2022a, 2022b, 2022). According to the Biomass Energy Centre, United Kingdom (Ahmed, 2018), biomass materials are categorized as shown in Figure 10.3. Recently, the continual and vigorous rising international oil prices and natural gas costs have given way to a rapid interest and increase in the commercialization of bioenergy production. Massive advances have been made in the production of methanol, ethanol, and

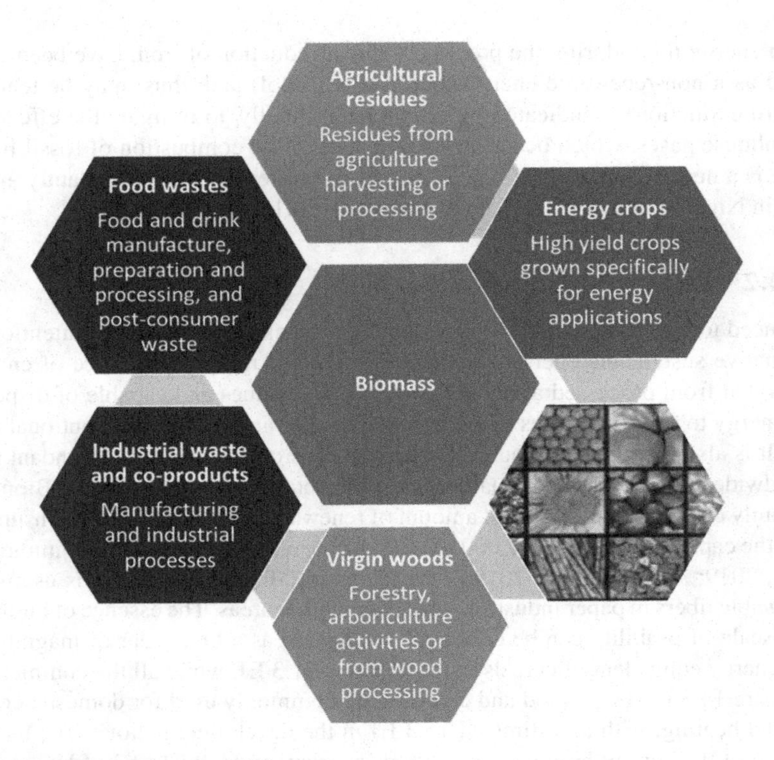

FIGURE 10.3 Categories of biomass materials.

Source: Grover, P. D., & Mishra, S. K. (1996). Regional wood energy development programme in Asia. Gcp/Ras/154/Net. RWEDP Report No. 23. In: *Proceeding of the International Workshop on Biomass Briqueeting*, New Delhi, India, 2–6 April, 1995. Out of courtesy (author's work).

biodiesel, and similarly, in the gasification biomass for biofuel production (Costello and Chum, 1998; Ikubanni et al., 2021; Balogun et al., 2021a, 2021b, 2022).

Since it is an organic reductant, agro-based energy sources can effectively serve as carbon input during the process of reducibility of iron ores or any iron-bearing materials. Biomass was used as a reductant in reducing limonite in the study of Wang et al. (2009). It was shown that biochar was more effective in reducing limonite and good magnetization. The reduction temperature was reduced to 100°C (a little above 650°C) (Wang et al., 2009). Brazil is known as the largest charcoal producer in the world, with over 2.3 million tons annually. Similarly, Brazil is concurrently the country in which bioenergy is most widely used in ironmaking processes. Pulverized charcoal that was used in BF operation in 2005 was above 0.19 million tons, and the pulverized charcoal injection ratio reached around 100–150 kg/t HM in Brazil. Increasingly, nearly 35% of the hot metal produced in 2007 from Brazil was obtained using charcoal with an internal volume of <300 m^3 as a reductant. Biochar produced by carbonizing unbroken babassu coconut at 1000°C could directly replace coke in BF operations (Emmerich and Luengo, 1994).

10.5 IRON RECOVERY PROCESS FOR DRI

The production process of sponge iron is classified based on the choice of the reducing agents. Irrespective of the nature of the reducing agent (coal-based or gas), an important factor is the accessibility and economic viability of those reducing agents. In oil-producing states, for example, there could be an abundance of natural gas, such as hydrogen, to their advantage. In this case, it will be economically friendly to use natural gas as a reducing agent. Additionally, countries like India, Nigeria and a host of others may wish to settle for coal-based reducing agents due to their abundant reserve, while developing countries, especially some of the SSA like Nigeria, can still opt for solid-solid reducing agents, especially biomass sourced to producing sponge iron due to large agricultural waste or residue recovered from their various agricultural assignments. Figure 10.4 indicates the iron recovery process for direct reduced iron. Iron ores or iron-bearing materials are considered primarily the most useful feedstock to produce direct reduced iron, followed by the reductant, whose objective is to enact the necessary energy for the production to take place. It was reported that the agglomeration of these most active materials makes reduction occur at comparatively low temperatures and with quality products. Also, it is economically efficient to agglomerate the fine iron-bearing materials and the reducing agent (carbonaceous materials) with the aid of suitable binders (organic or inorganic binders) and a reasonable amount of moisture for strength activation. The effect of carbon/hematite molar ratio on the

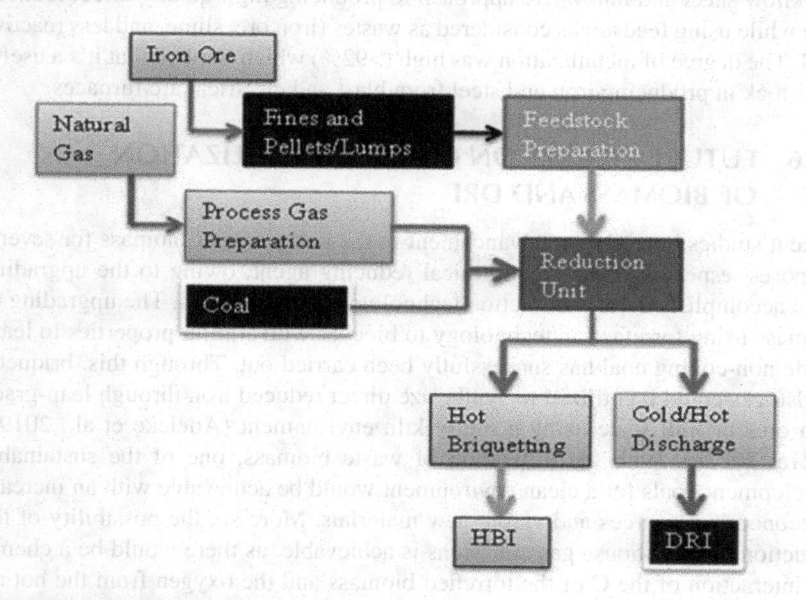

FIGURE 10.4 DRI recovery flow sheet.

Source: Anand, R. S., Kumar, P., & Paswan, D. D. N. (2016). Conversion of sponge iron from low grade iron ore and mill scale through simulating tunnel kiln condition. *IOSR Journal of Mechanical and Civil Engineering*, *13*(5), 49–54. https://doi.org/10.9790/1684-1305064954. Out of courtesy (author's work).

reduction capacity of iron ore-coal composite pellet reduced at 1250°C for 20 min in a multilayer bed rotary hearth furnace (laboratory scale) was studied by Mishra and Roy (2016). It was concluded that pellets with optimum carbon/hematite ratio (1.66) yielded the highest reduction improved carbon utilization, and productivity for the layer considered. The top layer exhibited maximum reduction at a relatively lower C/Fe_2O_3 molar ratio of <2.33 in the pellet, while the bottom layer surpassed the top layer reduction at a higher C/Fe_2O_3 molar ratio of >2.33. The relationship between the degree of reduction and degree of metallization points at non-isothermal kinetics was affected by heat and mass transfer in multilayer bed RHF. The crushing strength of the partially reduced pellet with optimum C/Fe_2O_3 molar ratio (1.66) indicated that they could be possibly used as a replacement feedstock for BF operations or in any other smelting reactor (Mishra and Roy, 2016; Somerville, 2016). Anand et al. (2016) carried out a possibility study of producing direct reduced iron from lean-grade iron ore/slime, mill scale and partly oxidized coal (Jhama coal). Like most research works, a composite mixture of iron-bearing materials, carbon source, binding agent, and the activator (water) was considered the most effective way of realizing green pellets before being charged into the furnace for proper reduction. However, in their work, this was not the case. The reduction of dried pellets was carried out without induration/firing at higher temperatures as compared to the conventional direct reduction process, yielding high energy efficient and cost-effective process. The work produced an optimized workflow sheet and innovative approach to producing high-quality direct reduced iron while using feedstocks considered as wastes (iron ore, slime, and less reactive coal. The degree of metallization was high (>92%) which implied that it's a useful feedstock in producing iron and steel from blast and electrical arc furnaces.

10.6 FUTURE STUDIES ON COMBINED UTILIZATION OF BIOMASS AND DRI

Recent studies have shown advancement in the utilization of biomass for several purposes, especially as a metallurgical reducing agent, owing to the upgrading level accomplished via torrefaction technology of the biomass. The upgrading of biomass using torrefaction technology to biochar with similar properties to lean-grade non-coking coal has successfully been carried out. Through this, briquette feedstocks could be utilized to synthesize direct reduced iron through lean-grade iron ores or mill scale using a rotary kiln environment (Adeleke et al., 2019b, 2021a, 2022b). With the utilization of waste biomass, one of the sustainable development goals for a clean environment would be achievable with an increase in economic resources and viable new materials. More so, the possibility of the reduction of greenhouse gas emissions is achievable, as there would be a chemical interaction of the C of the torrefied biomass and the oxygen from the hot air in the furnace, which will react with the lean-grade iron ore/mill scale. Hence, a continuous combustion of carbon present in the biomass would result in continued reduction. The production of DRI with the use of torrefied biomass as a reductant has attracted little attention. The utilization of torrefied biomass from various agro-wastes as reducing agents in DRI production should be explored, as some studies have found the partial usage biomass effective to have good reactivity (Adeleke et al., 2022). The utilization of biochar has also found good reactivity, except that the sintering speed and permeability as negatively affected. Further considerations on the use of pulverized torrefied biomass for making sponge iron, where it serves as a direct reducing agent, should be intensively explored.

10.7 SUMMARY

The direct reduction process is a major technique for obtaining sponge iron from iron-bearing materials. This has been thoroughly discussed in this chapter. Various iron-bearing materials were discussed, while several direct reduction processes were also emphasized. The use of various reductants, such as coal, gas, and hydrogen, was discussed. The potential usage of biomass as carbonaceous material to produce DRI (sponge iron) has been seen as a recent development to impact the world in terms of reduction in the emission of greenhouse gases. Agro (biomass) and industrial wastes are good substitutes, such as mill scale, and biomass is a good substitute for ore and coal in the production of DRI (sponge iron).

REFERENCES

Adebimpe, R. A. & Akande, J. M. (2011). Engineering economy analysis on the production of iron ore in Nigeria. *Geomaterials*, 1(1), 14–20.

Adeleke, A. A., Odusote, J. K., Ikubanni, P. P., Agboola, O. O., Balogun, A. O., & Lasode, O. A. (2021b). Tumbling strength and reactivity characteristics of hybrid fuel briquette of coal and biomass wastes blends. *Alexandria Engineering Journal*, 60(5), 4619–4625. https://doi.org/10.1016/j.aej.2021.03.069

Adeleke, A. A., Odusote, J. K., Ikubanni, P. P., Lasode, O. A., Malathi, M., & Paswan, D. (2021a). Essential basics on biomass torrefaction, densification and utilization, *International Journal of Energy Research*, 45(2), 1375–1395. https://doi.org/10.1002/er.5884

Adeleke, A. A., Odusote, J. K., Ikubanni, P. P., Olabisi, A. S., & Nzerem, P. (2022b). Briquetting of subbituminous coal and torrefied biomass using bentonite as inorganic binder. *Scientific Reports*, 12(1), 1–11. https://doi.org/10.1038/s41598-022-12685-5

Adeleke, A. A., Odusote, J. K., Lasode, O. A., Ikubanni, P. P., Malathi, M., & Paswan, D. (2019a). Mild pyrolytic treatment of Gmelina arborea for optimum energetic yields. *Cogent Engineering*, 6, 1–13. https://doi.org/10.1080/23311916.2019.1593073

Adeleke, A. A., Odusote, J. K., Lasode, O. A., Ikubanni, P. P., Malathi, M., & Paswan, D. (2019b). Densification of coal fines and mildly torrefied biomass into composite fuel using different organic binders. *Heliyon*, 5(e02160), 1–6. https://doi.org/10.1016/j.heliyon.2019.e02160

Adeleke, A. A., Odusote, J. K., Lasode, O. A., Ikubanni, P. P., Madhurai, M., & Paswan, D. (2022a). Evaluation of thermal decomposition characteristics and kinetic parameters of melina wood. *Biofuels*, 23(1), 117–123. https://doi.org/10.1080/17597269.2019.1646541

Adeleke, A. A., Ikubanni, P. P., Balogun, O. A., Okolie, Jude A., Nnodim, C. T., Olawale, A. O., & Okonkwo, C. J. (2022). Comparative analyses of lean grade coal and carbonized *Antiaris toxicaria* (biomass) for energy generation. *Petroleum and Coal*, 64(2), 339–349.

Adeleke, A. A., Ikubanni, P. P., Orhadahwe, T. A., Christopher, C. T., Akano, J. M., Agboola, O. O., Adegoke, S. O., Balogun A. O., & Ibikunle R. A. (2021c). Sustainability of multifaceted usage of biomass: A review. *Heliyon*, 7, e08025. https://doi.org/10.1016/j.heliyon.2021.e08025

Agbu, O. (2007). The iron and steel industry and nigeria's industrialization: exploring cooperation with Japan, pp. 1–146, https://cir.nii.ac.jp/crid/1130282271819441792

Ahmed, H. (2018). New trends in the application of carbon-bearing materials in blast furnace ironmaking. *Minerals*, 8(12), 1–20. https://doi.org/10.3390/min8120561

Ajimotokan, H. A., Ehindero, A. O., Ajao, K. S., Adeleke, A. A., Ikubanni, P. P., & Shuaib-Babata, Y. L. (2019). Combustion characteristics of fuel briquettes made from charcoal particles and sawdust agglomerates. *Scientific African*, 6, 1–7. https://doi.org/10.1016/j.sciaf.2019.e00202

Anand, R. S., Kumar, P., & Paswan, D. D. N. (2016). Conversion of sponge iron from low grade iron ore and mill scale through simulating tunnel kiln condition. *IOSR Journal of Mechanical and Civil Engineering*, 13(5), 49–54. https://doi.org/10.9790/1684-1305064954

Baba, A. A., Adekola, F., & Folashade, A. (2006). Quantitative leaching of a Nigerian iron ore in hydrochloric acid. *Journal of Applied Science and Environment Management*, 9(3) https://doi.org/10.4314/jasem.v9i3.17346

Balogun, A. O., Adeleke, A. A., Ikubanni, P. P., Adegoke, S. O., Alayat, A. M., & McDonald, A. G. (2022). Study on combustion characteristics and thermodynamic parameters of thermal degradation of guinea grass (Megathyrsus maximus) in N_2-pyrolytic and oxidative atmospheres. *Sustainability*, 14(1), 1–21. https://doi.org/10.3390/su14010112

Balogun, A. O., Adeleke, A. A., Ikubanni, P. P., Adegoke, S. O., Alayat, A. M., & McDonald, A. G. (2021a). Kinetics modeling, thermodynamics and thermal performance assessments of pyrolytic decomposition of Moringa oleifera husk and Delonix regia pod. *Scientific Reports*, 11(1), 1–12. https://doi.org/10.1038/s41598-021-93407-1

Balogun, A. O., Adeleke, A. A., Ikubanni, P. P., Adegoke, S. O., Alayat, A. M., & McDonald, A. G. (2021b). Physico-chemical characterization, thermal decomposition and kinetic modeling of Digitaria sanguinalis under nitrogen and air environments. *Case Studies in Thermal Engineering*, 26, 101138. https://doi.org/10.1016/j.csite.2021.101138

Bamalli, U. S., Moumouni, A., & Chaanda, M. S. (2011). A review of Nigerian metallic minerals for technological development. *Natural Resources*, 2(2), 87–91. https://doi.org/10.4236/nr.2011.22011

Béchara, R., Hamadeh, H., Mirgaux, O., & Patisson, F. (2018). Optimization of the iron ore direct reduction process through multiscale process modeling. *Materials*, 11(7), 1–18.

Bin Zuo, H., Hu, Z. W., Zhang, J. L., Li, J., & Liu, Z. J. (2013). Direct reduction of iron ore by biomass char. *International Journal of Minerals, Metallurgy and Materials*, 20(6), 514–521. https://doi.org/10.1007/s12613-013-0759-7

Chand, S., Paul, B., & Kumar, M. (2015). An overview of use of Linz-Donawitz (LD) steel slag in agriculture. *Current World Environment*, 10(3), 975–984.

Costello, R., & Chum, H. L. (1998). Biomass, bioenergy, and carbon management. In: *BioEnergy '98: Expanding BioEnergy Partnerships*, pp. 11–17.

Das, B., Prakash, S., Biswal, S. K., Reddy, P. S. R., & Misra, V. N. (2005). Studies on the beneficiation of Indian iron ore slimes using the flotation technique. In: *Centenary of Flotation Symposium*. Brisbane: AusIMM.

Das, B., Prakash, S., Mohapatra, B. K., Bhaumik, S. K., & Narasimhan, K. S. (1992). Beneficiation of iron ore slimes using hydrocyclone. *Mining, Metallurgy and Exploration*, 9, 101–103.

Dutta, S. K. & Sah, R. (2016). Direct reduced iron: Production. In: *Encyclopedia of iron, steel, their alloy*. Colas, R. et al. (Eds.), Boca Raton: CRC Press, Routledge Handbooks Online, pp. 1082–1108. https://doi.org/10.1081/e-eisa-120050996

Emmerich, F. G., & Luengo, C. A. (1994). Reduction of emissions from blast furnaces by using blends of coke and babassu charcoal. *Fuel*, 73(7), 1235–1236.

Fick, G., Mirgaux, O., Neau, P., & Patisson, F. (2014). Using biomass for pig iron production: A technical, environmental and economical assessment. *Waste and Biomass Valorization*, 5(1), 43–55. https://doi.org/10.1007/s12649-013-9223-1

Filippov, L. O., Severov, V. V., & Filippova, I. V. (2014). An overview of the beneficiation of iron ores via reverse cationic flotation. *International Journal of Mineral Processing*, 127, 62–69. https://doi.org/10.1016/j.minpro.2014.01.002

Goldring, D. C. (2003). Iron ore categorisation for the iron and steel industry. *Transactions of the Institutions of Mining and Metallurgy- Section B Applied Earth Science*, 112(1), 5–17. https://doi.org/10.1179/037174503225011162

Grobler, F. & Minnitt, R. C. A. (1999). The increasing role of direct reduced iron in global steelmaking. *Journal of South African Institution of Mining and Metallurgy*, 99(2), 111–116. https://hdl.handle.net/10520/AJA0038223X_2636

Grover, P. D., & Mishra, S. K. (1996). Regional Wood Energy Development Programme In Asia Gcp / Ras / 154 / Net. RWEDP Report No. 23. In: *Proceeding of the International Workshop on Biomass Briqueeting*, New Delhi, India, 2–6 April, 1995.

Hashem, N., Salah, B., & El-Hussiny, N. (2015). Reduction kinetics of Egyptian iron ore by non coking coal. *International Journal of Science and Engineering Research*, 6(3), 846–852. https://doi.org/10.15580/gjbms.2013.5.070113696

Huiting, S., & Forssberg, E. (2003). An overview of recovery of metals from slags. *Waste Management*, 23(10), 933–949. https://doi.org/10.1016/S0956-053X(02)00164-2

Ikubanni, P. P., Adeleke, A. A., Agboola, O. O., Adesina, O. S., Nnodim, C.T., Balogun, A. O., Okonkwo, C. J., & Olawale, A. O. (2021). Characterization of some commercially available Nigerian coals as carbonaceous material for direct reduced iron production. *Materials Today Proceedings*, 44(Part 1), 2849–2854. https://doi.org/10.1016/j.matpr.2020.12.1167

Ikubanni, P. P., Omololu, T., Ofoegbu, W., Omoworare, O., Adeleke, A. A., Agboola, O. O., & Olabamiji, T. S. (2019). Performance evaluation of briquette produced from a designed and fabricated piston-type briquetting machine. *International Journal of Engineering Research and Technology*, 12(8), 1227–1238.

Jena, S. K., Sahoo, H., Rath, S. S., Rao, D. S., Das, S. K., & Das, B. (2015). Characterization and processing of iron ore slimes for recovery of iron values. *Mineral Processing and Extractive Metallurgy Review*, 36(3), 174–182. https://doi.org/10.1080/08827508.2014.898300

Kirschen, M., Badr, K., & Pfeifer, H. (2011). Influence of direct reduced iron on the energy balance of the electric arc furnace in steel industry. *Energy*, 36(10), 6146–6155. https://doi.org/10.1016/j.energy.2011.07.050

Kumar, M., Jena, S., & Patel, S. K. (2008). Characterization of properties and reduction behavior of iron ores for application in sponge ironmaking. *Mineral Processing and Extractive Metallurgy Review*, 29(2), 118–129.. https://doi.org/10.1080/08827500701421896

Kumar, R. (2014). Carbonization study of non coking coals and characterisation of their properties for application in sponge iron making. Master's thesis. Department of Metallurgical and Materials Engineering, National Institute of Technology, Rourkela.

Kumar, R., & Mandre, N. R. (2017). Recovery of iron from iron ore slimes by selective flocculation. *The Journal of the Southern African Institute of Mining and Metallurgy*, 117, 397–400. https://doi.org/10.17159/2411-9717/2017/v117n4a12

Li, H. M., Li, L. X., Yang, X. Q., & Cheng, Y. B. (2015). Types and geological characteristics of iron deposits in China. *Journal of Asian Earth Science*, 103, 2–22 2015.

Marquardt, K., Rohrer, G. S., Morales, L. F. G. & Rybacki, E. (2014). The most frequent interfaces between olivine crystals," *92nd Annual Meeting DMG*, no. MIC-P04, pp. 232–233.

Mishra, S., & Roy, G. G. (2016). Effect of amount of carbon on the reduction efficiency of iron ore-coal composite pellets in multi-layer bed rotary hearth furnace (RHF). *Metallurgical and Materials Transactions B*, 47(4), 2347–2356. https://doi.org/10.1007/s11663-016-0666-1

Mitra Debnath, R., & Sebastian, V. J. (2014). Efficiency in the Indian iron and steel industry–an application of data envelopment analysis. *Journal of Advances in Management Research*, 11(1), 4–19. https://doi.org/10.1108/JAMR-01-2013-0005

Odesola, I. F., Samuel, E., & Olugasa, T. (2013). Coal development in Nigeria: Prospects Challenges. *International Journal of Engineering and Applied Sciences*, 4 (1), 64–73.

Odusote, J. K., & Muraina, H. O. (2017). Mechanical and combustion characteristics of oil palm biomass fuel briquette. *Journal of Engineering and Technology*, 8(1), 14–29.

Odusote, J. K., Adeleke, A. A., Ameenullahi, B. S., & Adediran, A. A. (2019). Preliminary characterisation of iron ores for steel making processes. *Procedia Manufacturing*, 35, 1123–1128. https://doi.org/10.1016/j.promfg.2019.07.020

Ogbesode, J. E., Ajide, O. O., Oluwole, O. O., & Ofi, O. (2023). Recent trends in the etchnologies of the direct reduction and smelting process of iron ore/iron oxide in the extraction of iron and steelmaking. In: *Iron ores and Iron oxide*. Kumar, B. (Ed.). IntechOpen Publisher, https://doi.org/10.5772/intechopen.1001158

Ohimain, E. I. (2013). The challenge of domestic iron and steel production in Nigeria. Greener. *Journal of Business and Management Studies*, 3, 231–240.

Olayebi, O. O. (2014). "Steel making experience in the use of Nigerian iron ore at the Delta Steel Company, Nigeria. *Journal of Chemical Engineering and Materials Science*, 5, 47–62. https://doi.org/10.5897/JCEMS2014.0166

Pal A., Samanta, A. N., & Ray, S. (2010). Hydrometallurgy treatment of iron ore slime for value addition. *Hydrometallurgy*, 105, 30–35. https://doi.org/10.1016/j.hydromet.2010.07.005

Pal, J. (2019). Innovative development on agglomeration of iron ore fines and iron oxide wastes. *Minerals Processing and Extrative Metallurgy Review*, 40(4), 248–264. https://doi.org/10.1080/08827508.2018.1518222

Qin, Y., Lv, X., Bai C., & Qiu, G. (2012). Waste heat recovery from blast furnace slag by chemical reactions. *Journal of the Minerals, Metals, and Materials Science*, 64, 997–1001. https://doi.org/10.1007/s11837-012-0392-3

Ramakgala, C., & Danha, G. (2019). A review of ironmaking by direct reduction processes: Quality requirements and sustainability. *Procedia Manufacturing*, 35, 242–245. https://doi.org/10.1016/j.promfg.2019.05.034

Roy, S. K., Nayak, D., & Rath, S. S. (2020). A review on the enrichment of iron values of low-grade Iron ore resources using reduction roasting-magnetic separation. *Powder Technology*, 367, 796–808. https://doi.org/10.1016/j.powtec.2020.04.047

Roy, S., & Das, A. (2013). Recovery of valuables from low-grade iron ore slime and reduction of waste volume by physical processing. *Particulates Science and Technology*, 31(3), 256–263. https://doi.org/10.1080/02726351.2012.716147

Sarkar, S., Sarkar, S., & Biswas, P. (2017). Effective utilization of iron ore slime, a mining waste as adsorbent for removal of Pb(II) and Hg(II). *Journal of Enviromental Chemical Engineering*, 5(1), 38–44. https://doi.org/1b 0.1016/j.jece.2016.11.015

Sinha, A., Biswas, P., Sarkar, S., Bora, U., & Purkait, M. K. (2020). Utilization of LD slag from steel industry for the preparation of MF membrane. *Journal of Environmental Management*, 259, 110060. https://doi.org/10.1016/j.jenvman.2019.110060

Somerville, M. A. (2016). The Strength and density of green and reduced briquettes made with iron ore and charcoal. *Journal of Sustainable Metallurgy*, 2(3), 228–238. https://doi.org/10.1007/s40831-016-0057-5

Umadevi, T., Kumar, M. G. S., Mahapatra, P. C., Babu, T. M., & Ranjan, M. (2009). Recycling of steel plant mill scale via iron ore pelletisation process. *Ironmaking and Steelmaking*, 36(6), 409–415.https://doi.org/10.1179/174328108X393795

Wang, G. C. (2016). Ferrous metal production and ferrous slags. In: The utilization of slag in civil infrastructure construction, pp. 9–33. https://doi.org/10.1016/B978-0-08-100381-7.00002-1

Wang, Y. B., Zhu, G. C., Chi, R. A., Zhao, Y. N., & Cheng, Z. (2009). An investigation on reduction and magnetization of limonite using biomass. *Chinese Journal of Process Engineering*, 9(3), 508–13.

WCST-World Crude Steel Production-The Largest Steel Producing Countries. https://worldsteel.org/media-centre/press-releases/2022/december-2021-crude-steel-production-and-2021-global-totals/ (accessed 23 November 2022).

White, L. (1929). Geography's part in the plant cost of iron and steel production at Pittsburgh, Chicago, and Birmingham. *Economic Geography*, 5(4), 327–334.

Wiedmann, T. O., Schandl, H., & Moran, D. (2015). The footprint of using metals: new metrics of consumption and productivity. *Environmental Economics and Policy Studies*, 17, 369–388. https://doi.org/10.1007/s10018–014–0085-y

Zachariah, S. M., Grohens, Y., Kalarikkal, N., & Thomas, S. (2022). Hybrid materials for electromagnetic shielding: A review. *Polymer Composites*, 43(5), 2507–2544. https://doi.org/10.1002/pc.26595

11 Sustainable Hybrid-Reinforced Metal Matrix Composites

A Review of Production and Characterization

P. P. Ikubanni
Landmark University, Omu-Aran, Nigeria

A. A. Adeleke
Nile University of Nigeria, Abuja, Nigeria

M. Oki
Greenfield Creations Ltd. Benue Close, Agbara, Ogun State, Nigeria

Harmanpreet Singh
Indian Institute of Technology Ropar, Rupnagar, India

11.1 INTRODUCTION

Choosing the right materials by materials engineers for applications is a grueling challenge in engineering. There is always a great relief when materials are carefully and properly selected to meet up with the designer's expectations, especially when such materials retain the desired properties throughout the period of utilization in a particular defined working environment. As stated by Nturanabo et al. (2019), it is nearly impossible to find a material (monolithic) that will have the necessary property profile for all technical applications. Therefore, it is important to develop materials with enhanced properties through the combination of at least two materials. This category of material is what is collectively called composite

DOI: 10.1201/9781003309123-11

materials (Mussatto et al., 2021; Adediran et al., 2022). A composite material is made up of a befittingly arranged mixture of at least two macro, micro, or nano constituents of different materials with a separating interface differing in form and chemical composition and is fundamentally non-dissolvable in each other (Smith and Hashemi, 2008). This implies that the identity of each constituent is maintained as they do not melt into each other. Composite helps to produce a new material with better properties (Prasad et al., 2014; Aigbodion and Ezema, 2020; Edoziuno et al., 2021; Kareem et al., 2021). The production of composite materials helps improve the strength-to-weight ratio of components that could be popularly utilized in the industrial and automobile sectors (Mavhungu et al., 2017; Ikele et al., 2022; Zemlianov et al., 2022). Additionally, the implementation of metal matrix composite (MMC) in the automotive, aerospace, agricultural, and mineral mining sectors has become the focus of numerous studies to lower weight and increase material performance with improved characteristics (Zakaria, 2014; Krupakara and Ravikumar, 2015; Bhoi et al., 2020). When two or more reinforcing particles are included in a single matrix, hybrid composites are created. The strengths or advantages of one particle type could complement the weaknesses or drawbacks of the other in the production of hybrid composites, which may involve the use of two or more types of particulates (Mahendra and Radha Krishna, 2010; Kareem et al., 2021; Suresh Kumar et al., 2022).

Aluminum and its alloys are popular engineering materials due to their great heat conductivity and lower weight, which makes them valuable in the automotive and other industrial engineering industries. In doing so, it will be possible to create a variety of high-performing components that may be used for numerous applications (Fatchurrohman et al., 2015). Additionally, because of their excellent corrosion resistance, low density, and superior electrical conductivity, aluminum and its alloys are being employed more and more (Kala et al., 2014). It is possible to significantly enhance some of the properties of aluminum by using reinforcement in ceramic forms, such as silicon carbide (SiC), silicon dioxide (SiO_2), alumina (Al_2O_3), titanium dioxide (TiB_2), and so on (Kumar et al., 2016; Aigbodion and Ezema, 2020; Kumar and Birru, 2018; Kanth et al., 2019; Gillani et al., 2022). MMCs have rapidly expanded as a result of the growing need for innovative materials in the aerospace and automotive industries (Allison and Cole, 1993; Narula et al., 1996; Alaneme et al., 2019; Edoziuno et al., 2021). The strength-to-weight ratio of aluminum matrix composites (AMCs) is high, with excellent formability for easier secondary processing (including hot extrusion, forging, and rolling), good wear resistance, reasonable corrosion resistance, high-temperature creep resistance, and fatigue strength (Macke et al., 2012; Kumar et al., 2016). For long-term applications where weight preservation is a vital characteristic, AMCs with high specific stiffness and high strength may be employed (Zakaria, 2014; Mavhungu et al., 2017; Suresh Kumar et al., 2022). In aluminum and aluminum alloy matrixes, ceramic elements such as Al_2O_3, SiC, magnesium oxide (MgO), and boron carbide (B_4C) have been widely employed as reinforcement. These materials are suited for use as reinforcement in composite matrixes due to their exceptional features, including high compressive strength and hardness, refractoriness, wear resistance, and so on (Lancaster et al., 2013). The most prevalent and widely used discontinuous reinforcements in metal matrix composites (MMC) in automobile industries, aerospace-structural applications, and casting applications are SiC, Al_2O_3, and B_4C as described by various researches (Cheng et al., 2007; Zakaria, 2014; Poornesh et al., 2016; Moona et al., 2018; Singh and Goyal, 2018; Dixit and Suhane, 2022).

In recent times, attention has gradually shifted from AMCs developed with traditional reinforcing particulates, including SiC, Al_2O_3, graphite, and so on, to the usage of agro and industrial waste derivatives (Prasad and Rama Krishna, 2012; Alaneme and Sanusi, 2015). The utilization of these agro- and industrial waste derivatives in a metal matrix helps in cost reduction, producing acceptable physical properties with applications in various areas, including thermal management, mild stress-bearing, and semi-structural areas (Alaneme and Olubambi, 2013). However, they are primarily applied in areas of relatively lesser strength and wear properties (Prasad and Shoba, 2014). Hence, agro and industrial wastes derived reinforcing materials have been used as complementing materials in addition to conventional synthetic reinforcing materials in various proportions. The outcome of these proportions is referred to as hybrid reinforcement composites (Escalera-Lozano et al., 2008; Alaneme and Adewale, 2013). Hybrid-reinforced composites of various combinations and compositions have been studied by various researchers. Some of the hybrid reinforcement combinations are alumina/rice husk ash/graphite (Alaneme and Sanusi, 2015), rice husk ash/alumina (Alaneme et al., 2013; Alaneme and Olubambi, 2013), alumina/graphite (Baradeswaran and Elaya Perumal, 2014), rice husk ash/silica carbide (Prasad et al., 2014), fly ash/silica carbide (Mahendra and Radha Krishna, 2010), and so on. However, the levels of performance obtainable using hybrid composite systems are dependent on the reinforcement type combination, percentage weight of the reinforcements in the composite, matrix-reinforcement wettability, processing technique adopted, and metallurgical characteristics of a matrix (Alaneme et al., 2014).

To assess the application potential of AMCs, the composites' susceptibility to corrosion is an important factor to be considered (Zakaria, 2014). Considerable works have evaluated the physico-mechanical and wear properties of Al MMCs. However, the corrosion tendency of Al MMCs in corrosive environment-assisted wear has yet to gain attention from researchers and, thus, is very scanty in the literature. From the literature, Zakaria (2014) revealed that there may be electrochemical, chemical, or physical interactions between the metal alloy and reinforcing particulates, which could result in accelerated corrosion of composites in certain media. However, other reinforcement particulates, at some percentage weight inclusions have been found to advance the corrosion resistance of MMCs made from Al-Mg-Si alloy matrix and groundnut shell ash and SiC particulates, such as at 6 and 10 wt.% of the hybrid reinforcements (Alaneme et al., 2015); Al6061 and TiC particulates at 2, 4, and 6 wt.% of TiC (Ananda Murthy and Singh, 2015); Al7075 and SiC particulates (Cheng et al., 2007); and Al-graphite combination with 4.5% and 8.5% magnesium addition (Wang et al., 2010). Various industrial environments have greatly influenced the properties of materials deployed in specific applications. Thus, researchers in engineering have been prompted to search far and wide for high-performance components with relatively low costs but environment-friendly composite materials. In addition, consumers' preferences and environmental demands for lighter, faster, and more durable automobiles, as well as machines, hand tools, building materials, and so on, have necessitated constant adjustments for higher standards in the market space. These and other considerations have ignited the research and development of novel and cutting-edge materials for definite applications in industrial and consumer product environments. Newly developed composite materials have found full implementations in the automobile industries, marine and aerospace industries, and so on, in which ceramic-reinforced materials have been broadly utilized for AMCs production by many researchers (Mavhungu et al., 2017; Sahu and Sahu, 2018). Currently, hybrid reinforcements composed of ceramic and green reinforcing materials, obtainable from agro-residues and industrial wastes, are being used as reinforcements in metal alloys for AMC development. The ashes produced from these agro-wastes are the major supplementary particulate fibers used as reinforcements by various groups with relative success.

Palm kernel shells (PKS) are agro-residues abundant in palm oil-producing regions of Nigeria (Figure 11.1). They are readily available in modified forms for usage as

FIGURE 11.1 PKS in a palm oil-producing area in Osun State, Nigeria.

reinforcement in metal matrixes. A lot of efforts have been geared toward the utilization of palm kernel shell ash (PKSA) for cement's partial replacement in construction works (Olutoge et al., 2012; Oti et al., 2015; Hardjasaputra et al., 2018). However, relatively few works in the literature have been dedicated to PKSA as a source of hybrid green reinforcement along with ceramics in AMCs' development. From the literature, the corrosion and wear characteristics of AMCs vary greatly. Therefore, it is necessary to examine the corrosion behaviors of newly developed AMCs from hybrid reinforcements such as PKSA and SiC. The use of chromate in conversion coatings for Al alloys has been identified to be carcinogenic (Oki et al., 2017). Literature showed that attention has been given to finding alternative routes for the use of chromate in conversion coating formation for AMCs to improve their corrosion resistance. Thus, novel conversion coating solutions and processes are used for the improvement of the corrosion resistance and good paint adhesion of the composites.

This study explored the utilization of PKSA as a green reinforcing material in AMC's development from aluminum alloy matrix and hybrid reinforcement materials for use in automobile, aerospace, marine, and sports industries. These are additional efforts in reducing the pollution generated through the disposal of PKSs and reducing the carbon footprints of industries through lower energy demands for processes and products developed therefrom. Agro-residues and industrial waste materials, when disposed of, can cause environmental pollution-related challenges. Therefore, the recycling of these wastes as reinforcement in MMCs because they contain ceramic materials that serve as hardeners and strengtheners to improve the properties of the matrix. Corn cub, PKSs, rice husks, coconut husks, and sugarcane bagasse, among others, are some of the materials that have been utilized in MMC production. Bodunrin et al. (2015) suggested that investigations should be performed on more agro-wastes. PKS are waste materials majorly utilized as aids in briquette making to serve as fuel for cooking. The PKSA obtained from the burning of the PKS has found utilization as a partial replacement for cement, which plays a significant function in concrete's strength and durability. This assists in alleviating the growing concern for scarce and cement's expensiveness (Olutoge et al., 2012; Oti et al., 2015). Moreover, it is believed that the ash particles of these burned shells have the potential for utilization as reinforcement materials. This present study investigates the characteristics of PKSA as a partial replacement for SiC as reinforcement in AMC production to improve its physical and mechanical properties for various applications. Thus, contributing to the achievement of the Industry, Innovation, and Infrastructure Goal 9 of sustainable development enunciated by the United Nations.

11.2 GENERAL OVERVIEW OF ALUMINUM AND ITS ALLOY

As reported by Nturanabo et al. (2019), aluminum is one of the prevalent metallic elements for extraction from the Earth's crust. Since the medieval period (fifth–fifteenth century) and the industrial revolution's early years, aluminum has been very important and has played important roles in domestic and manufacturing industries. One of the unique and broad materials for use as a matrix is aluminum and its alloys because they possess a low density-to-weight ratio, high specific modulus, strength properties, high thermal stability, high wear property, and good resistance to corrosion (Krupakara and Ravikumar, 2015; Ratna Kumar et al., 2017; Zemlianov et al., 2022). Furthermore, aluminum and its alloys possess high ductility and strength in cryogenic environments (Ratna Kumar et al., 2017). Aluminum alloys have found applications in engineering; hence, when materials engineers are searching for a preferred substitute to iron and steel, aluminum alloys have always been the choice due to their better physico-mechanical properties (Bodunrin et al., 2015; Nturanabo et al., 2019).

The progressive oxidation, which causes mild steel to rust away, is always resisted by aluminum by forming an inert aluminum oxide film when a freshly exposed aluminum surface combines with oxygen. The protective layer of aluminum will instantly reseal itself when scratched. The non-ferromagnetic property of aluminum makes it very important in the electrical and electronics industries. It displays excellent electrical and thermal conductivity. Aluminum corrosion product is non-toxic, a property that makes it useful in food and beverage packaging. The melting point is relatively low, which makes its casting process easier. The metal can be cast using any known method in the foundry and can be further worked upon into aluminum wire and roofing sheets for the building and allied industries (Davis, 2001).

11.2.1 Categories of Aluminum Alloys

Two major categories of aluminum alloys have been recognized: wrought and cast compositions. The basic mechanism involved in the development of this metal helps to classify each category. Owing to different phase solubilities, several alloys are responsive to heat treatment processes. These types of alloys (casting or wrought) are considered to be heat treatable. However, when it is wrought, the compositions are dependent on work-hardening via mechanical reduction. This involves the mixture of different annealing processes for the development of the property termed work-hardening. There are essentially some non-heat-treatable casting alloys. Hence, they mostly find applications in as-cast or thermally treated situations, which are unconnected to the precipitation or solution phenomena (Davis, 2001).

Different nomenclatures for wrought and cast alloys have been developed (Davis, 2001). The wrought alloys allow a four-digit system for producing a list of wrought composite families, while the casting compositions have a three-digit system designation and are accompanied by a decimal value. These wrought alloys compositions are in the 1xxx, 2xxx, 3xxx, 4xxx, 5xxx, 6xxx, 7xxx, 8xxx, and 9xxx series, while the cast alloys compositions are in the 1xx.x, 2xx.x, 3xx.x, 4xx.x, 5xx.x, 6xx.x, 7xx.x, 8xx.x, and 9xx.x series. All these different series have prime alloying elements. For instance, the alloying element in the 5000 series is magnesium, and in the 7000 series is zinc, while the alloying elements (magnesium and silicon) are inherent in the 6000 series. These different series are developed for various engineering applications. However, this study is concerned with the utilization of the aluminum wrought alloys category. Although some other series were mentioned, the 6xxx series was majorly considered in this study because they are heat treatable, have good formability, weldability, machinability, corrosion resistance, good dimensional stability, low thermal expansion coefficient, and excellent structural rigidity (Davis, 2001; Kumar et al., 2016).

11.2.2 Why Composite Materials?

During the past two decades, the utilization of composite materials has increased in popularity and steadily penetrated new markets. Contemporary composite materials are applicable to simple and sophisticated designs. It is important to state that other industries aside aerospace or aircraft industry have found composite materials commercially useful in recent years. Composite materials are weight, energy, and cost-saving materials. Some of the applications include engine cascades, curved fairing and fillets, replacement of cylinders, tubes, ducts, and so on. More so, the necessity for lighter construction materials, as well as more seismic-resistant structures, has made composite materials more sought after in new and advanced materials. The following are some of the advantages of composite materials over conventional ones:

1. Composite materials have tensile strength that is more than that of steel or aluminum. It is estimated to be between four and six times that of aluminum.
2. Composite materials have improved torsional stiffness and impact properties.
3. Composite materials have up to 60% of the ultimate tensile strength and, hence, a higher fatigue endurance limit than the base material.
4. The versatility of composite materials is more compared to that of metals. They can be fashioned to achieve performance needs and complex requirements.
5. Composite materials have improved appearance with smooth surfaces.

Recent engineering applications require cheap, lighter, and stronger materials. These find applications in vehicle and ancillary industries where superior strength-to-weight ratio materials are the most desired. The development of MMCs with hybrid reinforcements helps in achieving this.

11.3 OVERVIEW OF METAL MATRIX COMPOSITES

The blend of at least two components or phases that are chemically different on a microscopic measure, with distinguished boundary, and with ease of specification is referred to as a composite (Smith and Hashemi, 2008; Nturanabo et al., 2019; Bhoi et al., 2020). In a composite, the constituent materials are always present in reasonable proportions. In addition, the phases of the constituent should possess distinct properties that will make the composite properties different from the constituent's properties. The constituents can either be the matrix or reinforcements.

Matrix is available in larger quantities, and it is the continuous constituent. It is generally believed that during composite development, the properties of the matrix are the targets for improvement. Ceramic, metallic, or polymeric materials are the various forms of a matrix where each form of matrix material has distinct mechanical properties. The comparison among the three forms of matrix showed that metals have good ductility and strength and intermediate moduli; ceramics are strong, stiff but brittle materials, while polymers possess low strength and Young's moduli (Matthews et al., 2000; Macke et al., 2012). However, reinforcements help in the enhancement of the physico-mechanical properties of the matrix. Although, with some exceptions, reinforcement materials are most times harder, stiffer, and stronger than the matrix, such as in polymer matrixes.

The focus of this research work is on MMCs. MMCs are crafted out by the combination of at least two materials. One of these materials is usually a metal (matrix). Furthermore, when the matrix and reinforcing material are agglomerated to obtain MMC, the desirable properties of the MMCs are retained through the combination of the reinforcement strength with the matrix ductility (Kurumlu et al., 2012; Mussatto et al., 2021). Reinforcing constituents can take any form such as particle, fiber (short or continuous), or platelet, which can range in size from sub-micrometer to millimeter.

During the initial development of MMCs, continuous ceramic fibers, as well as single-crystal ceramic whiskers, have been the chosen reinforcements because remarkable strength and stiffness increases have been observed. However, particulate and discontinuous reinforcements have registered significant improvement on many fronts, especially when aluminum is used as the matrix for the production of the MMCs (Nturanabo et al., 2019). Aluminum or its alloys, among other metallic materials, have found utilization as base metal in MMCs development (Alaneme et al., 2013, 2018; Dinaharan et al., 2017; Edoziuno et al., 2021; Adediran et al., 2022). This is linked to aluminum or its alloy forming a penetrating network. The reinforcing materials are usually introduced into the matrix. The reinforcement materials that are usually utilized in a metal matrix are ceramics, which are often synthetic reinforcing materials. The ceramic reinforcing materials include boron carbide (B_4C), alumina (Al_2O_3), silicon carbide (SiC), and many more (Cheng et al., 2007; Loto and Babalola, 2017; Poornesh et al., 2016; Singh and Goyal, 2018). Ceramic particulates reinforced aluminum alloys have been found to be significantly useful for structural applications owing to their low density and high specific strength and stiffness (Falcon et al., 2011; Zakaria, 2014). It is worth stating that aluminum-based MMCs' resistance to corrosion is less than that of conventional alloys. This was a result of their modified behavior by phases and also due to grain mismatch at the interface of matrix-reinforcement, defects in manufacturing, internal stress, and other inherent factors (Cheng et al., 2007).

Recently, reinforcement materials have been obtained through the usage of industrial wastes such as fly ash (Bieniaś et al., 2003), red mud (Prasad, 2006), and the ash obtained from agricultural residues, such as bamboo leaf, rice husk, corncob, and many others on (Alaneme et al., 2013, 2018, 2019; Fatile et al., 2014; Prasad and Shoba, 2014). These reinforcing materials are obtained from industrial wastes and the agro-residues are usually mixed at relatively low proportions with ceramic reinforcing materials. Hence, hybrid reinforcing materials are developed with better mechanical properties for usage in the aerospace, automobile, and marine sectors.

11.4 CLASSIFICATION OF METAL MATRIX COMPOSITES

It is possible to classify MMCs based on their reinforcement's shape, type, and method. According to Surappa (2003) and Nturanabo et al. (2019), Al MMCs can be classified as follows:

11.4.1 Continuous Fiber-Reinforced MMCs

They fiber-reinforced MMCs comprise either monofilaments, coarse fibers or relatively fine fibers. Typical examples of fine continuous fibers are carbon, SiC, or Al_2O_3, having a size diameter that is below 20 μm. Before infiltrating into a composite formation, the fine, continuous fibers are either singly embedded or pre-woven. However, the range of shapes that the coarser fibers can produce is limited by their flexibility. Chemical vapor deposition (CVD) is used to deposit SiC or boron into a carbon fiber or tungsten core to produce monofilaments. Large-diameter fiber monofilaments have a diameter of between 100 and 150 μm.

11.4.2 Particle-Reinforced MMCs

This kind of MMCs can also be referred to as particulate-reinforced MMCs. They are majorly obtained using ceramic reinforcements such as borides (e.g., TiB_2), carbides (e.g., SiC), or oxides (e.g., Al_2O_3). These composite reinforcements are utilized with a below 5 aspect ratio and available for utilization with below 30% volume fractions. The composites are obtained from the amalgamation of the metal matrix with the ceramic powders at a pre-determined proportion. Solid-state sintering or liquid-metal methods (which are; squeeze casting, stir casting, and in situ processes) can be employed in the production of this kind of composite.

11.4.3 Whisker- and Short-Fiber-Reinforced MMCs

The aspect ratio of particle-reinforced MMCs is less than 5. However, the aspect ratio of the whisker- and short- fiber reinforcements of these composites is more than 5 and is not continuous. For instance, in the production of a piston, the foremost and leading reinforcement utilized is the short Al_2O_3. On the other hand, the powder metallurgy route or squeeze infiltration method can be used to produce whisker-reinforced composites into a preformed fiber, which is mostly developed to net/near-net shape. It is paramount to state that perceived health hazards have restricted the use of whiskers as reinforcement materials (Nturanabo et al., 2019).

11.4.4 Hybrid-Reinforced MMCs

All other aforementioned MMCs mostly contain a single reinforcement material inclusion within the matrix metal. However, hybrid MMCs comprise at least two different reinforcement types. For instance, hybrid MMCs are produced when particle and whisker, fiber and particle, etc., are combined as reinforcement materials in a matrix. However, hybrid MMCs of ceramic and agro-residues or industrial waste reinforcement materials have been used to produce advanced composite materials (Boopathi et al., 2013; Alaneme et al., 2018; Kareem et al., 2021; Ikele et al., 2022). With the discovery of carbon nanotubes (CNT), they enjoy current usage as reinforcements in MMC products. It was reported that the addition of CNT into a matrix alloy in composite production improved the corrosion and mechanical characteristics of the MMCs produced (Ratna Kumar et al., 2017).

11.5 METHODS OF AMCs PRODUCTION

A significant issue with composite materials processing is to have evenly distributed reinforcement phases in order to obtain a composite with a relatively uniform microstructure. Proper selection of reinforcements, as well as the methods of processing AMCs, are very crucial factors in ensuring the achievement of desired property combinations (Fatile et al., 2014; Bhoi et al., 2020). The technology of production adopted can determine the extent of heterogeneity of the developed composite. Hence, it is important to examine the different classifications of the composite production processes. The primary processes for AMC based on an industrial scale can be classified as described by Mavhungu et al. (2017) as follows.

11.5.1 Liquid State Processes (LSP)

LSP comprises different processing techniques. These techniques are highlighted in this sub-section.

11.5.1.1 Stir-Casting Technology

This technology is a significant liquid state process that helps to distribute particulate reinforcement materials into the matrix melt via mechanical stirring. Stir casting is a cheap method for creating AMCs. According to Mavhungu et al. (2017), this technique can result in composites with a reinforcement volume fraction of approximately 30%. This technology is easy, fast, efficient, and flexible. The matrix structure can be more easily controlled by varying the stirrer blades and cooling times (Parswajinan et al., 2018b). Smith and Hashemi (2008) reported that due to particles settling during solidification, reinforcing particles are segregated using the stir-casting route. As a result, the relative density, solidification rate, particle wetting state, and mixing strength all have a role in the final solid particle dispersion. A newly developed and improved stir-casting method called double stir casting, a two-step mixing technique, has been employed to yield a homogeneous distribution of the particles by some researchers in various studies (Alaneme et al., 2014; Prasad et al., 2014; Alaneme and Sanusi, 2015, Kareem et al., 2021; Adediran et al., 2022). The procedure entails heating matrix metal (aluminum) above the liquidus temperature. After, a semi-solid state is attained by allowing the melt to cool. This state, which lies between the liquidus and the solidus temperature allows preheated reinforcement materials ease of incorporation into the matrix. At this elevated temperature, the slurry is methodically mixed with a mechanical stirrer. The strength of this technology is its capacity to disrupt the gas barrier surrounding the particle surface, which prevents wetting between the molten metal and reinforcing particle. Therefore, compared to traditional stirring, this method results in a more homogeneous microstructure (Kala et al., 2014). In the production of AMCs, researchers have utilized different mechanical stirring speeds between 200 and 600 rpm (Alaneme and Bodunrin, 2011; Krishna and Xavior, 2014; Prasad et al., 2014; Alaneme and Sanusi, 2015; Jeykrishnan et al., 2016; Poornesh et al., 2016; Rajesh and Santosh, 2017). The final micrograph of the manufactured AMCs is impacted by the stirring speed. According to Raei et al. (2016), the addition of the reinforcing particles is equally distributed throughout the matrix, and the product's mechanical properties improve with increased stirring speed and time. Figure 11.2 depicts a simple schematic layout of the stir-casting procedure.

Mavhungu et al. (2017) reported that some researchers are of the view that the stir-casting method provides a better and wider range of shapes with a larger size of products up to 500 kg when compared with other methods. Moreover, it prevents or alienates severe damage as well as reinforcement fractures, which are associated with squeeze casting and powder metallurgy processes (Mavhungu

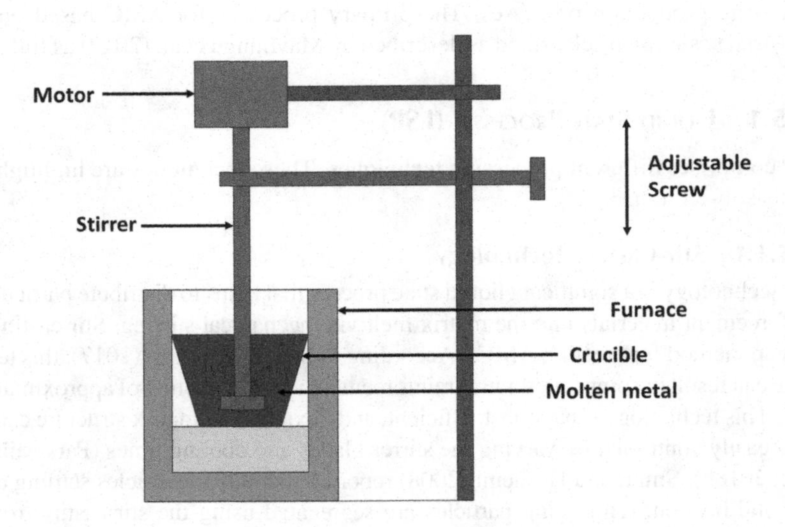

FIGURE 11.2 Schematic view of the stir-casting process.

Source: Haque, S., Bharti, P. K., & Ansari, A. H. (2014). Mechanical and machining properties analysis of Al6061-Cu-reinforced SiC_P metal matrix composite. *Journal of Minerals and Materials Characterization and Engineering*, *2*(1), 54–60. https://doi.org/10.4236/jmmce.2014.21009. Used under CC BY-4.0.

et al., 2017). Due to other competing technologies' comparatively higher costs, stir casting is AMC's preferred production method. Furthermore, the stir-casting method produces products of higher quality (both mechanically and physically) if the process variables, such as mixing time and stirring speed, are carefully monitored and optimized. For more enhanced physical and mechanical qualities of AMCs, the stirring speed, stirring time, feed rate, impeller diameter, position of the impeller, and stirrer blade angle should all be carefully evaluated.

11.5.1.2 Squeeze Casting

One of the attractive processing methods by which AMCs can be produced is squeeze casting. Squeeze casting combines casting and forging processes through the aid of high-applied pressure during molten metal solidification. The produced AMCs exhibit better mechanical properties as porosity and shrinkage cavities are reduced, and reinforcement segregation is eliminated (Dhanashekar and Senthil Kumar, 2014). To produce quality AMCs products, this process utilizes high-applied pressure with the least turbulence and low die-filling velocity (Aweda and Adeyemi, 2009). Dhanashekar and Senthil Kumar (2014) highlighted two different forms of squeeze casting, which are direct and indirect squeeze casting. The technique utilized in direct squeeze casting during solidification involves the pressure application on the whole liquid-metal surface via a punch to produce full-density castings. However, with the indirect method, metal is injected into the die cavity in the indirect squeeze casting by a smaller diameter piston (Yue and Chadwick, 1996). The reinforcement particle dispersion plays a crucial part in obtaining the material's anticipated properties. More so, the wettability improvement of the reinforcing particulates and high pressures are applied during casting (Seyed Reihani, 2006; Sukumaran et al., 2008). These are achieved when this technology is employed.

A few of the advantages of squeeze casting are: (i) the fluidity of the alloy is not critical, (ii) gas porosity or shrinkage is drastically reduced from the parts produced, (iii) feeders or risers are not needed, (iv) there is no metal wastage, and (v) good mechanical properties are obtainable due to grain size reduction (Dhanashekar and Senthil Kumar, 2014; Kopeliovich, 2012a). However, casting pressure, preheat/die temperature, casting/preform, and solidification rate greatly impact the properties of the composites produced (Hwu et al., 1996). This process is limited to a preformed shape of up to 2 cm in height, and the process is moderately expensive (Mavhungu et al., 2017). The illustrative picture of the process is revealed in Figure 11.3.

11.5.1.3 Spray Deposition

This method involves generating and developing the blend of the droplets of metal matrix together with particles of ceramics, which are then sprayed on a removable substrate. There is usually a quick matrix solidification with reinforcing particle inclusion and declination in reaction time between the base alloy and the reinforcing materials, as displayed in Figure 11.4. However, there might be the formation of residual porosity and the high expense of the utilization of inert gas, including considerable material wastage throughout the process. This spray-deposition method has been utilized by Cheng et al. (2007).

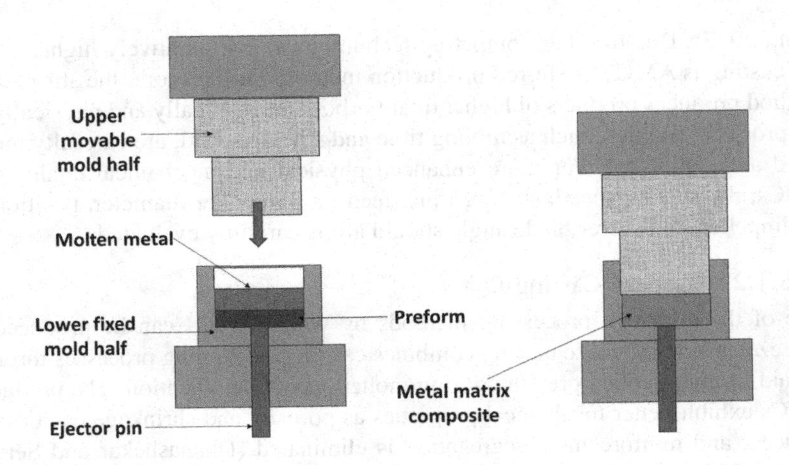

FIGURE 11.3 Illustrative display of squeeze casting technique.

Source: https://www.substech.com/dokuwiki/doku.php?id=liquid_state_fabrication_of_metal_matrix_composites. Out of courtesy (author's work).

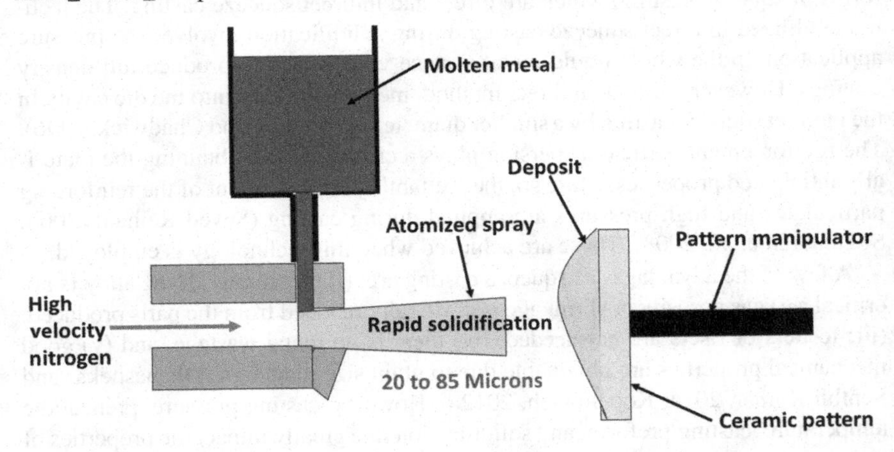

FIGURE 11.4 Spray-deposition technique.

Source: Li, B.T., Lavernia, E.J., Lin, Y., Chen, F., & Zhang, L. (2016). Spray forming of MMCs. In: Reference Module in Materials Science and Materials Engineering, p. 6. https://doi.org/10.1016/B978-0-12-803581-8.03884-4. Reproduced with permission.

11.5.1.4 In Situ (Reactive) Processing

This process dodges the necessity for intermediate development of the reinforcement. The reinforcements are developed through the metal matrix in situ reaction in a single phase (Kopeliovich, 2012b). There are usually very clean matrix-reinforcement interfaces that ensure improved bonding and wetting between the reinforcing particulates and the base metal. Figure 11.5 is a typical illustration of the mechanism of the in situ (reactive) process of developing MMCs.

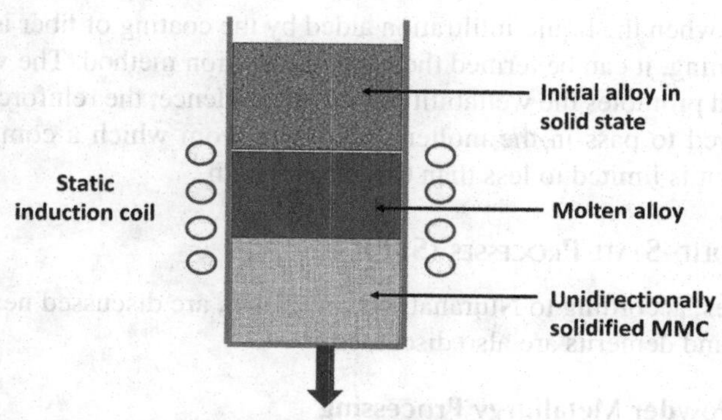

FIGURE 11.5 In situ processing technique.

Source: https://www.substech.com/dokuwiki/doku.php?id=in-situ_fabrication_of_metal_matrix_composites. Out of courtesy (author's work).

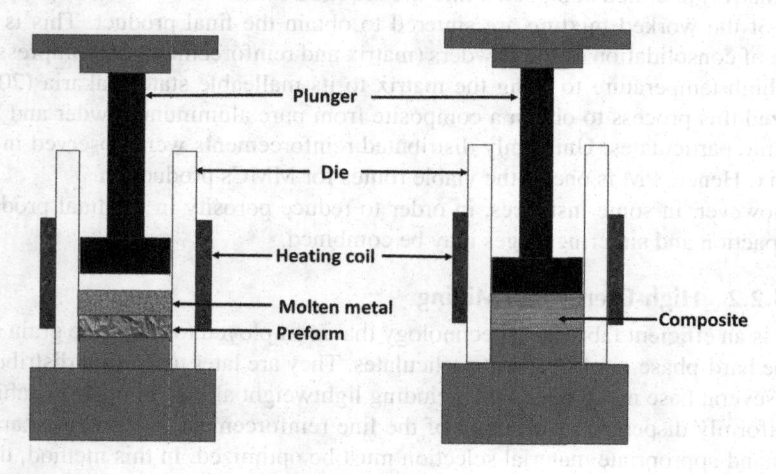

FIGURE 11.6 Liquid-infiltration technique.

Source: Sree Manu, K. M., Ajay Raag, L., Rajan, T. P. D., Gupta, M., & Pai, B. C. (2016). Liquid metal infiltration processing of metallic composites: A critical review. *Metallurgical and Materials Transactions B*, *47*, 2799–2819. Out of courtesy (author's work).

11.5.1.5 Liquid Infiltration

Liquid infiltration is mostly used when creating composites with long-fiber. In this method, ceramic filaments are organized into files, and mats develop inside preforms where liquid metal is injected either through pressure, vacuum, or gravity, depending on the approach. When the process is done using a vacuum, it is termed vacuum infiltration, and when done under gravity, it is termed pressure-less infiltration. The schematic procedure for the production of MMCs using this method is shown in Figure 11.6.

However, when the liquid infiltration aided by the coating of fiber is combined with hot forming, it can be termed the vapor deposition method. The vapor deposition method promotes the wettability of the fiber. Hence, the reinforcing materials are allowed to pass in the molten metal bath from which a composite wire comes out that is limited to less than 0.5 cm in length.

11.5.2 Solid-State Processes (SSP)

The processes, according to Nturanabo et al. (2019), are discussed next. Some of their merits and demerits are also discussed.

11.5.2.1 Powder Metallurgy Processing

From the raw materials to the finished product, there are three steps in powder metallurgy (PM) technology (Figure 11.7). The foremost stage involves the preparation of the metal powder required to form the metal matrix. The second stage is concerned with the proper mixing of powders involved (reinforcement particles and matrix) and then compacted into the required forms. In the third stage, powders of the worked mixture are sintered to obtain the final product. This is the stage of consolidation of the powders (matrix and reinforcement) by compression at a high temperature to bring the matrix to its malleable state. Zakaria (2014) utilized this process to obtain a composite from pure aluminum powder and SiC ceramic particulates. Uniformly distributed reinforcements were observed in the matrix. Hence, PM is one of the viable routes for MMCs production.

However, in some instances, in order to reduce porosity in the final product, compaction and sintering stages may be combined.

11.5.2.2 High-Energy Ball Milling

This is an efficient fabrication technology that is employed to lower the grain size of the hard-phase reinforcement particulates. They are later uniformly distributed into several base metal matrixes, including lightweight alloys. In other to achieve a uniformly dispersed distribution of the fine reinforcement, the process parameters and appropriate material selection must be optimized. In this method, there is transfer of mechanical energy through high-frequency and high-energy balls exerting high-impact forces on the material being produced. This technique found application in high-density nanostructured MMC powder development. This is best suited for improved mechanical properties in synthesized high-density nanostructured MMC powders. Ansari and Cho (2016), as presented in Figure 11.8,

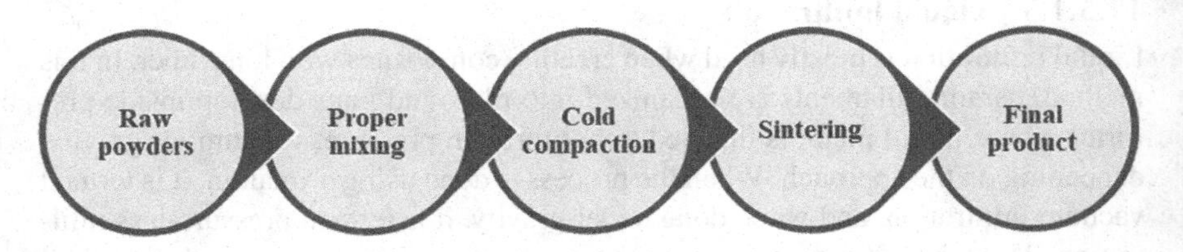

FIGURE 11.7 Step-wise process involved using the PM route.

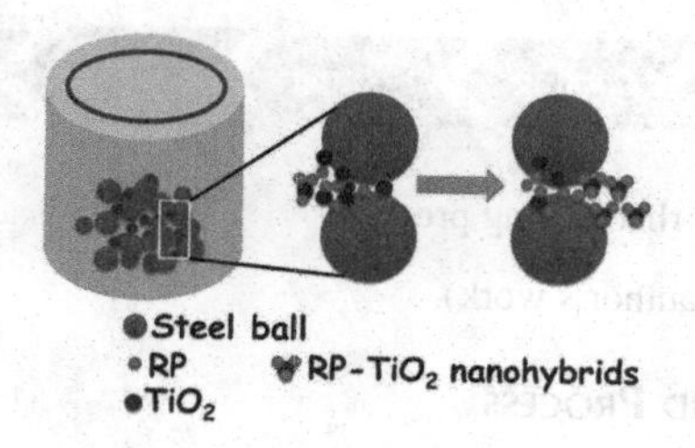

FIGURE 11.8 High-energy ball milling processing mechanism of nanohybrids of red phosphorus and TiO_2.

Source: Ansari, S. A., & Cho, M. H. (2016). Highly visible light responsive, narrow band gap TiO_2 nanoparticles modified by elemental red phosphorus for photocatalysis and photo-electrochemical applications. *Scientific Reports*, *6*(January), 1–10. https://doi.org/10.1038/srep25405. Used under CC BY-4.0.

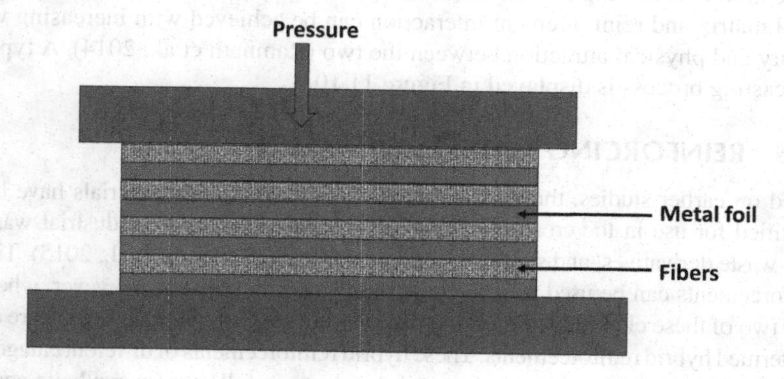

FIGURE 11.9 Schematic of diffusion bonding.

Source: https://www.substech.com/dokuwiki/doku.php?id=solid_state_fabrication_of_metal_matrix_composites. Out of courtesy (author's work).

proposed the mechanism for synthesizing the nanohybrids of red phosphorus and TiO_2 via the technology of high-energy ball milling. Generally, the production of nanosized materials on a large scale is achievable via a high-energy ball milling process. This is due to its ability to effectively grind large particles into nano sizes based on the developed forces through the mechanical action of grinding.

11.5.2.3 Diffusion Bonding

The diffusion bonding technique requires the utilization of both high temperature and pressure temperature for the tight packing of the matrix and fiber (Figure 11.9). It is a very good method for composite production with high mechanical characteristics (Kopeliovich, 2012c). However, high-energy intensities such as high pressures and temperatures are required for this process. The process is cumbersome since it is unsuitable for complex shape fabrication. For complex shape development, diffusion bonding is unsuitable. The process is more often utilized during Ti-based composite production with the reinforcement of fiber materials.

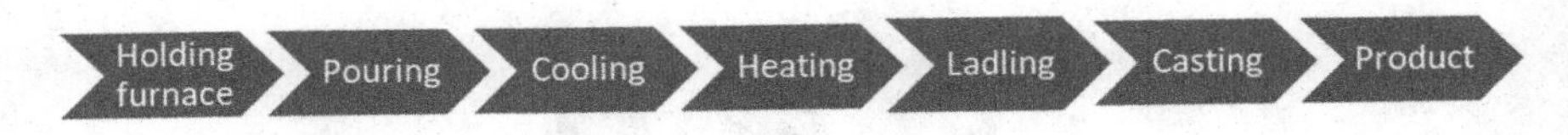

FIGURE 11.10 Typical rheocasting process.

Source: Out of courtesy (author's work).

11.5.3 Liquid-Solid Process

This process is otherwise referred to as compo-casting or rheocasting. It is an alternative technique to the stir-casting method, where semi-solid-state alloys are fused with reinforcing particulates. In compo-casting or rheocasting, the extremely viscous semi-solid slurries of the metal alloy receive the preheated reinforcing particulates incorporated via aggressive agitation. As a result, the pro-eutectic phase in the alloy slurry traps the reinforcing particles, preventing segregation. The continuous stirring action of the slurry makes it less viscous; hence, mutual metal matrix and reinforcement interaction can be achieved with increasing wettability and physical attraction between the two (Ramnath et al., 2014). A typical rheocasting process is displayed in Figure 11.10.

11.6 REINFORCING MATERIALS IN AMCs

Based on earlier studies, three major categories of reinforcing materials have been identified for use in the creation of AMCs. The three classes are industrial wastes, agro-waste derivatives, and synthetic ceramic particles (Bodunrin et al., 2015). These reinforcements can be used as a single reinforcement in a matrix; however, when at least two of these classified reinforcing particles are employed together, they are usually termed hybrid reinforcements. These hybrid reinforcements of different categories will be further considered under three sub-headings as follows: (a) synthetic ceramics hybrid-reinforced AMCs, (b) synthetic ceramic-agro-waste derivative reinforced AMCs, and (c) industrial waste-synthetic ceramic-reinforced AMCs. In considering the application of reinforcing materials in a matrix, parameters, including hardness, reinforcement type, dispersion in the matrix, modulus of elasticity, shape, and so on, are important to be examined (Das et al., 2014; Bodunrin et al., 2015). The ultimate qualities of the hybrid-reinforced AMCs are determined by the particular properties of reinforcement chosen, as well as the specific properties of the matrix alloy. Additionally, the final characteristics are influenced by the processing method used for producing the AMCs from the matrix and reinforcement (Alaneme and Aluko, 2012; Alaneme and Bodunrin, 2013; Alaneme et al., 2013; Casati and Vedani, 2014).

11.6.1 Synthetic Ceramics Hybrid-Reinforced AMCs

Silica (SiO_2), boron carbide (B_4C), silicon carbide (SiC), carbon nanotubes (CNT), alumina (Al_2O_3), graphite (Gr), and tungsten carbide (WC) are examples of synthetic ceramic reinforcing materials. All these synthetic ceramic materials have been studied when used as reinforcement in a matrix. However, the use of SiC and Al_2O_3 has taken pride of place when compared with other reinforcing particulates. As reported by Alaneme and Bodunrin (2013) and Bodunrin et al. (2015), there has been an improvement in the strength and specific stiffness of the AMCs with reinforcing particles of SiC or Al_2O_3 against unreinforced alloys. However, the improvement's occurrence had negative include on the reinforced AMC's fracture toughness and ductility. To prevent material failures when under service stress, such materials should be ductile and tough to resist fracture. Unfortunately, previous studies have not been consistent on the corrosion property and performance of the AMCs developed from these synthetic ceramic materials reinforcements (Alaneme and Bodunrin, 2011; Alaneme et al., 2013). Consequently, researchers have devised means to optimize the performance of these AMCs by introducing hybrid synthetic materials as reinforcement.

Parswajinan et al. (2018a) utilized a hybrid combination of SiC and CNT in the manufacturing of AMCs using stir-casting technology based on four different compositions. When the hardness and impact behaviors of the AMCs were examined, the inclusion of the hybrid reinforcing materials improved the hardness and the impact strength of the AMCs developed. The corrosion behavior of the hybrid-reinforced composite was not considered in the study. Through the use of stir casting, Haider et al. (2015) investigated the impact of employing hard ceramics, SiC and Al_2O_3, as reinforcing components. According to the findings, the produced composite's tensile strength, hardness, and impact strength all increased. In a different study, SiC/Gr and SiC/Al_2O_3 hybrid reinforcements' effects on the wear characteristics of Al 6061-T6 hybrid composites made using the friction stir process were examined by Deravaju et al. (2013). According to reports, the hybrid AMCs had a more even distribution of reinforcing components than the metal matrix, which increased the hardness and wear resistance of the manufactured AMCs. Additionally, it was found that composites made of SiC and Al_2O_3 are harder than composites made of SiC and Gr. The fact that Al_2O_3 is harder than Gr made this possible. However, the composites with Gr components had superior solid lubricating effects than Al_2O_3 when the wear behavior of the composites was considered.

Rajesh and Kaleemulla (2016) examined the influence of SiC and alumina on the mechanical behavior of AMCs in a different investigation. For the development of a comparable hybrid composite, Al 7075 alloy was reinforced with SiC and alumina using the stir-casting technique. In comparison to the unreinforced alloy, the hybrid-reinforced (SiC and Al_2O_3) samples showed improved mechanical properties. In a related study, Ramnath et al. (2014) found that the produced AMCs displayed greater degrees of hardness and impact toughness than the unreinforced alloy. The reinforcing elements used to produce the AMCs were Al_2O_3 and B_4C. However, compared to the unreinforced alloy, the hybrid composites' tensile strength and flexural characteristics were lower. Padmavathi and Ramakrishnan (2014) investigated the tribological behavior of hybrid composite produced where SiC and multiwall carbon nanotube serve as the reinforcing materials. The stir-casting method was adopted in the study. The results revealed that there was a lesser friction coefficient and wear rate when the composite was tested under mild wear conditions. Thus, the tribological behavior of the reinforced samples was superior to the aluminum alloy. However, when the composites were subjected to severe wear conditions, the frictional coefficient and wear rates were observed to be higher. Therefore, it can be concluded that the force of load applied to the composites influenced the tribological behavior of the composites. More so, the composite's hardness value increased with increased reinforcement's percent volume. Ravindran et al. (2013) explored the usage of solid lubricants such as graphite and synthetic ceramic (SiC) as reinforcing particles for the production of aluminum hybrid nanocomposites. The graphite content was varied up to 10 wt.% while the SiC content was fixed at 5 wt.%. The hybrid composites' mechanical characteristics and microstructural behavior were also studied. It was claimed that as the reinforcement was added, the hardness value, wear resistance, and tensile strength increased. Comparing the hybrid composites to the single reinforced composite, better mechanical properties were found. According to a few of the previously published studies, adding reinforcing elements to a metal matrix significantly improves the mechanical features (such as tensile strength, hardness, and tribological behavior) of composites. In comparison to monolithic reinforced samples, hybrid reinforcements displayed improved mechanical and tribological behavior.

11.6.2 Industrial Waste-Synthetic Ceramic-Reinforced AMCs

Fly ash and red mud are two types of industrial waste that can be obtained from power plants and the aluminum sector (Boopathi et al., 2013; Krupakara and Ravikumar, 2015; Dinaharan et al., 2016; Reddy and Srinivas, 2018) and are cited as being suited for use as AMC reinforcement materials. A few research works have been reported on the usage of red mud as reinforcing material in AMCs. For instance, Krupakara and Ravikumar (2015) investigated the corrosion behavior influence of red mud reinforcement particles on the Al6061 metal matrix in seawater. It was reported that the corrosion rate in seawater reduced as exposure time increased for the composites. This may have resulted from adherent corrosion products formed initially at cathodic sites on the composite. Such adherent scales usually spread over the whole surface and effectively block the AMC from the environment with prolonged immersion time. According to Dinaharan et al. (2016), coal combustion in thermal energy plants results in the generation of waste products known as fly ash. It is found in massive amounts in the universe and causes land pollution. As reported by Bodunrin et al. (2015), two types of fly ash exist in Canada: class F, obtainable through bituminous coal combustion, and class C, obtainable through the combustion of sub-bituminous coal and lignite. The low chemical composition of calcium oxide (CaO) in the class F fly ash makes it preferable over class C for MMCs development. The main chemical constituents present in fly ash are silica and alumina, with trace composition of magnesium oxide and calcium oxide. The attractive benefits of using fly ash as a reinforcing material are because of its low density and cost (Bieniaś et al., 2003). Fly ash (FA) inclusion into a liquid matrix using gravity casting has been said to produce unevenly dispersed FA particulates in the composites. However, with the usage of squeeze casting, better homogenization and compatibility were recorded between the base metal and the particulates of FA (Bieniaś et al., 2003). Due to varying properties observed when FA is employed as a single reinforcement in an Al matrix, low weight and poor wettability are experienced; hence, hybrid reinforcement of synthetic ceramic reinforcement and fly ash have been investigated.

Boopathi et al. (2013) developed hybrid MMCs using aluminum, SiC, and FA. The compositions were made and the stir-casting method was adopted for the fabrication of the AMCs. It was reported that the density of the composites decreased while hardness increased. Superior yield and tensile strength were reported for the hybrid-reinforced composites compared to the unreinforced. However, there was a decline in the elongation of the reinforced composites. The effects of wear and corrosion were not investigated in the study. FA particle was combined with CNT as reinforcements in the production of AMCs using Al6061 (matrix). The production was done using the stir-casting method. The prepared samples were mechanically characterized to obtain tensile, flexural, hardness, and impact tests. The findings discovered hardness value increment with a decrease in FA content. More so, the impact strengths of the samples were observed to remain unchanged with any variation (Parswajinan et al., 2018a). Senapati and Mohanta (2016) evaluated the mechanical characteristics of hybrid-reinforced AMCs where SiC and FA are the reinforcing materials. The mechanical properties tested were better than the unreinforced sample for the various combinations of the hybrid reinforcements used. Additionally, with the weight fraction increment of SiC and FA reinforcements in aluminum matrix hybrid composites, significant improvements were discovered for ultimate tensile strength, hardness, and wear characteristics (Reddy and Srinivas, 2018). Escalera-Lozano et al. (2007) studied the corrosion behavior of hybrid-reinforced composites, which were produced using FA and SiC_p as reinforcing particulates in recycled aluminum. SiC_p is the silicon carbide particulate. The $MgAl_2O_4$ was developed in situ through the matrix-reinforcement reaction of Al matrix and FA. Al-8Si-15Mg and Al-3Si-15Mg were the two aluminum alloy metal matrixes used in the study. The matrix-reinforcements interface wettability yielded improvement with the inclusion of Mg and Si in high quantities. SiC and uncalcined FA were employed in the strengthening of the first matrix, whereas SiC and calcined FA were used to impregnate the second matrix. For the corrosion tests, the hybrid composite with SiC and calcined FA reinforcements did not corrode, while the composite with SiC and uncalcined FA reinforcing materials was corroded within the first month. Pitting corrosion, as well as the release of gray powders and trans-granular and inter-granular cracks, were observed. The suitable better result of the Al-3Si-15Mg/SiC_p/$MgAl_2O_4$ was credited to the positive inhibition of the Al_4C_3 phase caused by the availability of SiO_2 in the FA as well as the absence of carbon in the FA due to calcination. Carbon would have constituted a veritable cathode site for the corrosion reactions.

11.6.3 Synthetic Ceramic-Agro-Waste Derivative Reinforced AMCs

Agro-waste derivatives have been discovered as complementing reinforcements when used along with synthetic reinforcement. These agro-waste derivatives have helped in the production of a new class of hybrid AMCs. The derivatives of agro-waste, when used in AMCs synthesis, offer some uniqueness such as low cost, low density, easy accessibility, and reduction in environmental pollution (Bodunrin et al., 2015). The most important end-product of agro-waste for usage as a complementing reinforcing material is the ash obtained after burning.

The ash is usually conditioned in a muffle furnace for the removal of carbonaceous substances and volatile matter from the ash, which can affect its usability as reinforcement (Alaneme et al., 2015, 2018). Bodunrin et al. (2015) highlighted that the usage of agro-waste ash for developing AMCs on a commercial scale, has a very promising prospect because most developing countries have inadequate synthetic reinforcements. Moreover, in developing nations with adequate synthetic reinforcing materials, the cost of purchase is relatively high.

Some of the agro-wastes that have been used in previous studies include palm kernel shell ash (PKSA), rice husk ash (RHA), sugarcane bagasse ash (SBA), bamboo leaf ash (BLA), groundnut shell ash (GSA), corn cob ash (CCA), maize stalk ash (MSA), and so on (Ikubanni et al., 2020a, 2020b, 2021a, 2021b, 2021c, 2021d, 2022a; Alaneme et al., 2013; Fatile et al., 2014; Alaneme et al., 2015, 2018; Dinaharan et al., 2017; Akinlabi et al., 2019). From all the previous studies, it was noted that the properties of the AMCs were more than the unreinforced alloys. However, AMCs products with synthetic reinforcements gave superior properties to the agro-wastes ash-reinforced AMCs. Hence, it is important to harness the strengths of both agro-waste derivatives and synthetic reinforcements to obtain superior hybrid AMCs.

In their 2018 study, Alaneme et al. examined the microstructure, mechanical, and fracture properties of a hybrid AMC reinforced with SiC and GSA. The SiC (synthetic) and GSA (agro-waste derivative) mixed at different ratios constituted 6–10 wt.% of the reinforcement using Al6063 alloy as the matrix. The technique (two-step stir casting) was employed to develop the hybrid composite. The microstructure revealed the continuous phase of aluminum having a relatively uniform reinforcing particulate dispersion. The particle dispersion level was said to be easily discernible from their micrographs. In comparison to the unreinforced alloy, the single and hybrid-reinforced composites had higher hardness values. However, for samples with GSA as complementing reinforcement, the hardness value was lowered as the GSA wt.% increased in the composite. Moreover, the percentage volume of the oxides of Al, Si, Ca, K_2, and Mg contained in the GSA constituent caused a minor drop in the specific and ultimate tensile strengths of the composites for both 6 and 10 wt.%-reinforced Al-based composites. On the contrary, there was an increment in the fracture toughness as the GSA content increased, while there was a slight improvement of the percentage elongation which is invariant to GSA content increases. It was further reported that GSA content was more favorable to influencing the mechanical properties of hybrid composites when compared to RHA and BLA-reinforced composites. Chawla and Shen (2001), as conveyed by Alaneme et al. (2018), stated that direct and indirect mechanisms are responsible for the strength enhancement of AMCs. While the indirect strengthening mechanism involves an increase in dislocation density at the matrix-particle interfaces attributable to processing thermal mismatch, the direct strengthening mechanism involves load transfer between the base matrix and the reinforcing material.

The corrosion-resistant properties of Al6063 reinforced with GSA and SiC and immersed into 3.5% NaCl and 0.3M H_2SO_4 were examined in a different study by Alaneme et al. (2015). The corrosion performance of the produced composites was studied using weight loss and corrosion rate measurement. The developed composites were more corrosive in 0.3M H_2SO_4 solution but less corrosive in

3.5% NaCl solution. In the 0.3M H_2SO_4 solution, it was documented that superior corrosion resistance was observed with the composite tagged with Al-Mg-Si/6wt.% GSA-SiC having GSA and SiC weight ratios (1:3 and 3:1). The corrosion resistance improvement of some samples was linked to the availability of the combined action of silica and alumina in the GSA reinforcement particle. This was possible because the silica quantity in the developed composites assisted in preventing Al_4C_3 phase formation. It was further reported that the formation of the Al_4C_3 phase has a negative impact on the composites' resistance to corrosion. More so, the presence of Al_2O_3 in the composites is said to help in reducing the opportunity of experiencing galvanic corrosion occurrence between the SiC particles and the base alloy; hence, GSA-SiC has the tendency of reducing the formation of micro-galvanic cells, which could bring about Al-Mg-Si alloy dissolution.

To develop hybrid AMCs, Al-Mg-Si alloy was strengthened with SiC and BLA (Alaneme et al., 2013). The SiC particles were combined with BLA at various weight ratios, yielding 10 wt.% of the reinforcement that was used in the two-step stir-casting procedure to create the composite. The mechanical properties, corrosion behavior, and microstructure were investigated to determine the performance of the MMCs (Alaneme et al., 2013). The result showed that increased BLA content caused a reduction in the properties examined. Also, the hybrid MMCs showed superior fracture toughness than the singly reinforced Al-10 wt.% SiC composite. When the hybrid composites were submerged in 3.5 wt.% NaCl solution, the 2 and 3 wt.% BLA exhibited better corrosion resistance than the singly reinforced Al-10 wt.% SiC composite, and when submerged in 0.3M H_2SO_4, the single reinforced composite outperformed the hybrid composite.

Prasad et al. (2014) studied composites produced with the same quantity (between 2% and 8%) of RHA and SiC using the stir-casting protocol. The microstructural examination showed that there was a homogenous spreading of the reinforcing particles in the base metal. An increment in the amount of reinforcement increases the hardness, yield strength, and ultimate tensile strength values, while an inverse relationship was observed for percentage elongation and thermal coefficient of expansion (CTE) while reinforcing particulate content increased. Due to the differing coefficients of thermal expansion of the materials, the strengthening process was attributed to thermal mismatch. This CTE mismatch usually generates dislocation density increment with increasing content of the reinforcements. More so, the hybrid composites' response to aging at 155°C was observed to be faster than the monolithic alloy. Owing to the higher composites' dislocation density, the observed time required for the hybrid composites to reach maximum hardness was higher compared to the monolithic alloy. Therefore, agricultural waste derivatives utilization as reinforcing material complements for the synthesis of hybrid MMCs can cause the ductility, together with the composites' fracture toughness, to improve without any significant strength reduction (Bodunrin et al., 2015).

Despite the potential of agricultural waste alternatives for the reduction in cost and sustaining the mechanical properties performance, few investigators have considered the impact of agro-waste reinforcing particulates introduced into AMCs on the tribo-corrosion performance when utilized in applications involving corrosion and wear attacks. These derivatives are identified to potentially have the ability to suppress the Al_4C_3 phase based on the composition of the agro-wastes having silica of above 50% in their composition.

11.7 PHYSICO-MECHANICAL CHARACTERISTICS OF AMCs

The physical and mechanical properties of some formulated AMCs are abundantly reported in the literature. The selection was built on using different reinforcement materials which could influence the physico-mechanical characteristics of AMCs developed. The physical properties of developed composites that are normally investigated are density and porosity, while the mechanical properties that have been examined are tensile strength (yield strength (YS), ultimate tensile strength (UTS), and % elongation (%El.)), hardness, impact strength, fracture toughness, wear, and tribological behavior, among others.

Density is a physical property that is usually investigated when composites are produced. Experimental density is mostly evaluated using the principle of Archimedes, while the theoretical density is calculated with the utilization of the rule of mixtures based on the reinforcement weight fraction (Boopathi et al., 2013; Alaneme et al., 2013). Porosity is a vital physical property to be determined because it influences the composites' mechanical properties. It is usually calculated by using the values obtained for the experimental and theoretical densities of a sample (Alaneme et al., 2013; Prasad et al., 2014). According to Haque et al. (2014), to accurately, rapidly, and economically determine the deformation of materials, a hardness test should be carried out. This could be done using either the Brinell hardness testing machine, as used in the work of Haque et al. (2014) or the Rockwell hardness machine, as used in the study of Krishnappa et al. (2018) or the Vickers hardness machine, as adopted by Narayan and Rajeshkannan (2017) and Alaneme and Sanusi (2015). More so, the toughness of materials signifying the material's ability of energy absorbance when deforming plastically can be obtained through an impact test (Haque et al., 2014). The Izod impact testing machine was used by Haque et al. (2014), while the Charpy impact testing machine was used by Jeykrishnan et al. (2016) and Singh and Goyal (2018).

In the determination of the mechanical properties of AMCs, the tensile behavior of the produced composites is usually investigated. To measure the sample's UTS, YS, and %El., the Universal Testing Machine (UTM) is typically employed. When the produced samples are subjected to rubbing action repeatedly, there is always a slow and progressive irrecoverable loss of materials. This situation is termed wear, which results in the repair and, ultimately, the replacement of the worn-out parts (Xavier and Suresh, 2016). As reported by Xavier and Suresh (2016), MMCs wear resistance is dependent on some factors, including particle size, volume fraction, shape, and reinforcement material distribution. Wear studies have been carried out on AMCs using different wear equipment such as tribometer (Krishnappa et al., 2018); pin-on-disk test apparatus (Padmavathi and Ramakrishnan, 2014; Xavier and Suresh, 2016); and Taber abrasion wear testing machine (Alaneme and Sanusi, 2015). A few studies on the physico-mechanical properties results obtained after characterizing various MMCs produced are shown in Table 11.1.

TABLE 11.1
Physico-Mechanical Characterizations of MMCs

Matrix/Reinforcement Composition	Density (g/cm³)/ % Porosity	Hardness	UTS (MPa)	%El.	Reference
Al7075/5% SiC	2.810 (0.71%)	64 HRB	–	–	Surya and Gugulothu (2022)
Al7075/10%SiC	2.824 (0.87%)	67 HRB	–	–	
Al7075/15%SiC	2.839 (0.68%)	70 HRB	–	–	
EN AW 6061	–	53.3 HV	146	15.7	Petrović et al. (2022)
EN AW 6061/5% Al_2O_3/1% WSA	–	84.14 HV	185	10.8	
EN AW 6061/5% Al_2O_3/2% WSA	–	90.51 HV	189	9.6	
EN AW 6061/5% Al_2O_3/3% WSA	–	96.05 HV	193	8.9	
Al7075	2.780 (1.07%)	138.36 HV	87	6.4	Sambathkumar et al. (2021)
95% Al7075/5% RM	2.786 (1.64%)	175.56 HV	326	12.0	
90% Al7075/10% RM	2.795 (2.10%)	178.68 HV	272	10.2	
85% Al7075/15% RM	2.799 (2.73%)	181.18 HV	223	10.0	
Al	2.66	22.5 BHN	90.8	–	Ahamed et al. (2016)
Al/3% RHA	2.52	24.7 BHN	95.7	–	
Al/6% RHA	2.50	29.8 BHN	112.9	–	
Al/9% RHA	2.47	33.6 BHN	115.3	–	
Al6063	2.63 (1.5%)	–	–	–	Edoziuno et al. (2021)
Al6063/2.5% PKS	2.62 (1.84%)	–	–	–	
Al6063/5% PKS	2.61 (1.97%)	–	–	–	
Al6063/7.5% PKS	2.60 (2.26%)	–	–	–	
Al6063/10% PKS	2.59 (2.41%)	–	–	–	
Al6063/12.5% PKS	2.57 (3.31%)	–	–	–	
Al6063/15% PKS	2.55 (3.81%)	–	–	–	
Al6063Al6063/2% SiC	2.64 (2.06%)	73.02 BHN	116.09	7.6	Ikubanni et al. (2021)
Al6063/2% SiC/2% PKSA	2.65 (2.17%)	75.29 BHN	124.47	7.2	
Al6063/2% SiC/4% PKSA	2.61 (2.39%)	76.83 BHN	125.57	6.7	
Al6063/2% SiC/6% PKSA	2.58 (2.08%)	77.71 BHN	126.42	6.3	
Al6063/2% SiC/8% PKSA	2.54 (2.15%)	78.74 BHN	127.66	6.0	
	2.50 (2.28%)	80.55 BHN	128.17	5.8	

– WSA (walnut shell ash); RM (red mud); RHA (rice husk ash); PKS (palm kernel shell); PKSA (palm kernel shell ash)

11.8 ALUMINUM ALLOYS AND AMC UTILIZATION IN SEVERAL INDUSTRIES

The importance of aluminum-based materials in industries cannot be overemphasized. They have found applications in the automotive industries, aerospace and aircraft industries, marine transport industries, building and construction industries, and other new areas such as electronic and thermal management, packaging and containerization, electrical transmission, and sports industries.

11.8.1 Automotive Industry

Research and development have engineered the discoveries of light materials that offer better performance than existing materials (Mavhungu et al., 2017). AMCs have proven to be appropriate replacements for both steel and aluminum alloys in various automobile systems and components. These achievements were possible through the use of lower-density aluminum composites as replacements for carbon steel. The automotive industry has attracted many discoveries in which Al-based matrixes are being utilized in vehicle components such as valve covers, crank cases, control arms, suspension links, and unibody construction (all aluminum bodies), and so on (Nturanabo et al., 2019). The usage of AMCs as a replacement for steel is for weight saving without compromising the efficient output of the components. The most important automobile parts that AMCs have successfully produced are pistons and cylinder liners, main bearings, connecting rods, suspension, and brakes. Others include the rocker arm, engine cradle, piston ring groove, valve train, piston rod, and pin (Mavhungu et al., 2017; Sahu and Sahu, 2018). Figure 11.11 shows some of the applications of MMCs in automobile industries in producing some automobile parts.

11.8.2 Aerospace and Aircraft Industry

In these industries, aluminum-based components have enhanced the potential for mankind to travel into space and the world. Aluminum-based components have found utilization in some aircraft parts. The usage of AMCs consistently over other metals has been attributed to their weight reduction, thermal management, and mechanical stability (Nturanabo et al., 2019). Some of its main applications in aerospace and aircraft industries include jet engine blades, satellite solar reflectors, and missile fins. Others include metal mirror optics, aircraft electrical doors, precision components, wing panels, struts, etc. (Sahu and Sahu, 2018). Figure 11.12 illustrates some MMCs applications in airspace and aircraft industries.

11.8.3 Marine Transport

The implementation of AMCs has made it possible to expand the size and speed of boats, ships, ferries, and yachts while also enhancing their seaworthiness, safety, reliability, fuel efficiency, and cost-effectiveness. About 35%–45% weight

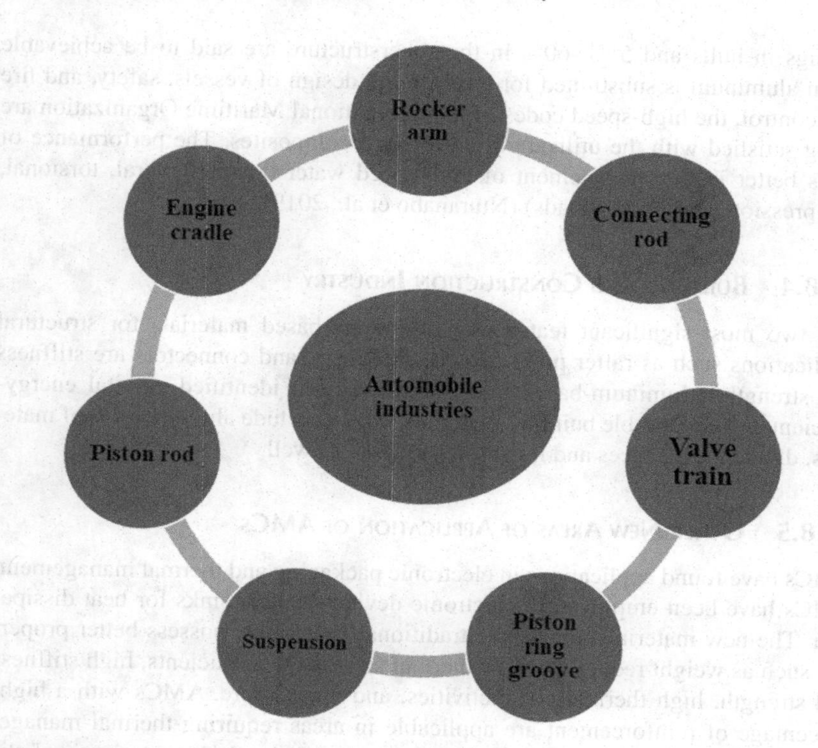

FIGURE 11.11 Automobile parts produced using MMCs.

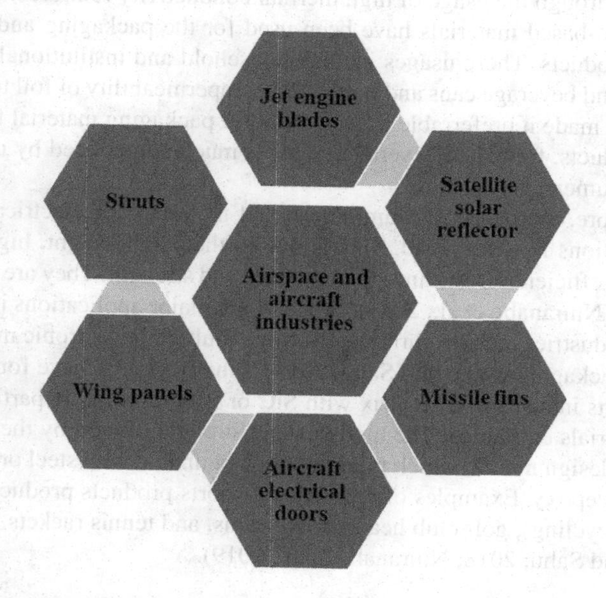

FIGURE 11.12 MMCs applications in airspace and aircraft industries.

savings in hulls and 55%–60% in the superstructure are said to be achievable when aluminum is substituted for steel. In the design of vessels, safety, and fire risk control, the high-speed codes of the International Maritime Organization are being satisfied with the utilization of Al-based composites. The performance of Al is better in the management of high-speed water travel (flexural, torsional, compression, and impact loads) (Nturanabo et al., 2019).

11.8.4 Building and Construction Industry

The two most significant features of aluminum-based materials for structural applications such as rafter poles, trusses, fasteners, and connectors are stiffness and strength. Aluminum-based materials have been identified as vital energy-efficient and sustainable building materials, which include alu-zinc for roof materials, dividers for offices and it is used for doors as well.

11.8.5 Other New Areas of Application of AMCs

AMCs have found applications in electronic packaging and thermal management. AMCs have been employed in electronic devices as heat sinks for heat dissipation. The new materials, unlike the traditional heat sinks, possess better properties such as weight reductions, low thermal expansion coefficients, high stiffness and strength, high thermal conductivities, and many more. AMCs with a high-percentage of reinforcement are applicable in areas requiring thermal management. This will help to continually improve the thermal characteristics of the composites through the usage of high thermal conductivity reinforcements. More so, aluminum-based materials have been used for the packaging and containerization of products. These usages include household and institutional foils, food containers, and beverage cans and bottles. The impermeability of foil to air, water, and light has made it preferrable as a material for packaging material for pharmaceutical products, food, and beverages. This is much appreciated by manufacturers and consumers.

Furthermore, the usage of aluminum-based materials for electrical transmission applications is widespread. This is due to their lightweight, high electrical conductivity efficiency, resistance to corrosion, and strength. They are also easy to be recycled (Nturanabo et al., 2019). Some of the major applications in electrical/electronic industries include current collectors, multichip electronic modules, and electronic packaging (Sahu and Sahu, 2018). Finally, AMCs have found applications in sports industries. Al matrix with SiC or B_4C reinforcing particles are the typical materials employed. The modulus and strength offered by these materials have better design merits, which might not be obtainable with steel or composites from carbon/epoxy. Examples of recreational/sports products produced are bicycle frames (cycling), golf club heads, wheel rims, and tennis rackets, among others (Sahu and Sahu, 2018; Nturanabo et al., 2019).

TABLE 11.2
AMCs Application in Various Industries (Mavhungu et al., 2017; Sahu and Sahu, 2018; Nturanabo et al., 2019)

Industry	Applications
Automotive	Engine cradle, pumps, valve train, crankshaft main bearing, piston rod, piston pins, cylinder blocks, sprocket, piston ring groove, disk brake rotors, rocker arm, piston crown, connecting rod, brake rotor, struts, drive shaft, propeller shaft, wheels, calipers brake disk, gears, pulley, cylinder heads, gearbox, gears, and rear drum
Aerospace/aircraft	Jet engine blades, space shuttle mid-fuselage tubular, wing panel, metal mirror optics, missile fins, aircraft electrical A.C. doors, satellite solar reflector, precision components, struts
Electronic Applications	Electric heat sink, electronic packaging, current collectors, multichip electronic module
Sports	Golf club heads, bicycle frames, tennis racket skis, wheel rims
Packaging	Beverage cans and bottles, food containers

The summarized overview of the AMCs application in various industries is presented in Table 11.2.

11.9 SUSTAINABLE UTILIZATION OF PALM KERNEL SHELL AND ITS ASH AS REINFORCEMENT IN MMCs PRODUCTION

Palm kernels (nuts) are obtainable from palm trees (Okoroigwe et al., 2014). Palm kernel is obtained when the oil-containing mesocarp has been removed (Anyaoha et al., 2018). Following the removal of the kernel (nut) from the hard shell (endocarp), PKS is obtained (Ikumapayi and Akinlabi, 2018). The lignocellulose presence in PKS is high. Palm oil processing leads to the generation of PKS, which becomes agricultural residue (Adebayo, 2012). The PKS and the nut can be separated through either manual (traditional) or mechanical (mechanized) techniques (Ibikunle et al., 2018). PKS is an important biomass source. PKS is widely available in nations that produce palm oil, such as Indonesia, Malaysia, Nigeria, Brazil, Sri Lanka, and others (Okoroigwe et al., 2014; Mortimer et al., 2018; Imoisili et al., 2020).

PKS has found usage in numerous applications due to its high content of solid wastes from processing palm oil. PKS is used in the fueling of boilers for the production of power as a result of their high-energy value (Demirbas, 2004) and as biofuel for domestic usage in loose or densified forms. It has been used for the reduction of ash deposition to increase heat and power production potential (Yin et al., 2008). PKS exhibits a propensity for bed agglomeration during fluidized bed combustion as a result of its high alkali concentration (Ninduangdee and Kuprianov, 2014). Researchers have identified additional uses for PKS, such as biofertilizer, biomass, water purification, concrete reinforcing additives, innovative materials development, energy storage, and supercapacitor electrode (Ibhadode and Dagwa, 2008; Mohamad et al., 2011; Yusuf and Jimoh, 2011; Fono-Tamo and Koya, 2013; Osei, 2013; Yacob et al., 2013; Fono-Tamo et al., 2014; Mgbemena et al., 2014; Okoroigwe et al., 2014; Obi, 2015; Rahman et al., 2016; Anyika et al., 2017; Ishola et al., 2017; García et al., 2018; Ikumapayi and Akinlabi, 2018; Misnon et al., 2019). The complete burning of PKS usually leads to ash derivation. The ash is typically produced when PKS has burned completely. Palm kernel shell ash (PKSA) is potentially utilized as a partial alternative for cement in building materials (Olutoge et al., 2012; Oti et al., 2015). Additionally, it might be employed in the creation of lightweight composite materials. About 5% PKSA by weight of solid wastes is often left after the complete burning of PKS (Olutoge et al., 2012). Ashes can be produced in a variety of colors, from whitish-gray to darker shades, depending on the quantity of carbon content present in the ash. Tangchirapat et al. (2007) opined that the indiscriminate disposal of PKSA could result in an environmental challenge. Some research efforts have been geared toward bringing solutions to this issue of good utilization of PKSA.

In a geopolymer composite, Hardjasaputra et al. (2018) examined the impact of employing both PKSA and RHA. The outcomes were contrasted with those of concrete made of Portland cement. When added to Portland cement concrete, the effects of using PKSA and RHA as partial cement replacements were determined to be negligible. The study illustrated PKSA development, but it did not consider how temperature affected the composition of PKSA produced under various temperature regimes. Olutoge et al. (2012) investigated the feasibility of partially substituting PKSA for cement in the production of concrete. It was determined that its use at lower volumes aided in the concretes' ability to contain less cement. Numerous researchers have investigated the utilization of PKSA as reinforcement in AMC (Oladele and Okoro, 2016; Akinlabi et al., 2019; Aigbodion and Ezema, 2020). In the fabrication of AMC utilizing the alloy A356, PKSA was employed as reinforcement in the form of nanoparticles (Aigbodion and Ezema, 2020). The influence of employing PKSA as a reinforcing material on the mechanical characteristics of recycled aluminum alloy derived from the cylinder of a car engine block was examined by Oladele and Okoro (2016). Before being used as reinforcement in the as-cast aluminum alloy, some of the ash derived from PKS was treated with a NaOH solution, while the other half was not. The results obtained demonstrated that PKSA addition is appropriate for the manufacture of AMCs, which have applications in automotive components. In order to produce AMC using the friction stir processing technique, Akinlabi et al. (2019) also used PKSA as reinforcement. A novel composite was produced by embedding the PKSA in an aluminum substrate surface. Other recent studies on PKS without the ash and SiC were carried out by Edoziuno *et al.* (2021), while Ikele et al. (2022) combine PKSA with SiC as reinforcements to form AMCs. Hybrid-reinforced AMCs were synthesized in the works of Ikubanni et al. (2021a, 2021b) using PKSA and SiC at various ratios with the maximum volume weight percentages of the reinforcements to be 10 wt.%. The combined effects of these reinforcing materials enhanced the tribo-mechanical properties of the synthesized composites.

11.10 MICROSTRUCTURAL CHARACTERIZATION OF MMCs WITH PKS AND PKSA AS REINFORCEMENT

The investigation of the internal morphology of any developed MMC through the utilization of microstructural equipment is very important. This assists in determining how well the particles are dispersed in the matrix metal after processing and solidification. The particles of the reinforcements could be either homogenously distributed or agglomerated at various positions. The form of the reinforcement dispersions in the matrix has different impacts on the mechanical and tribological behaviors of the developed MMCs. The morphologies of the unreinforced Al6063 alloy and reinforced alloy with 10 wt.% PKS were examined in a study by Edoziuno et al. (2021), as revealed in Figure 11.13. A fracture-like smooth surface was reported for the matrix alloy, whereas the presence of the PKS particulates resulted in toughening mechanisms. PKS particulates, as reinforcement, in the matrix alloy revealed homogeneous dispersion. The result of the EDS showed the presence of some of the elements detected by the equipment used. In a related study by Ikubanni et al. (2021), both the unreinforced alloy and reinforced MMCs, with hybrid compositions of SiC and PKSA at various weight percentages, were examined. Smooth surfaces without voids were reported for the unreinforced alloy. There was an even dispersion of reinforcements used in the metal matrix because the appropriate processing technique (double stir casting), as well as processing parameters, were utilized. The homogenous distribution of the particulates and the absence of cracks or voids were also linked to the matrix-reinforcement wettability. Generally, surface tension is always broken between

(a)

Element	At. %	Wt. %
Al	83.22	82.44
C	9.45	4.17
Sn	0.67	2.93
Ag	0.67	2.66
Fe	0.72	1.48
Ti	0.66	1.15
Si	1.02	1.06
Ca	0.62	0.91
Mg	0.91	0.81
S	0.63	0.74
K	0.50	0.71
P	0.55	0.62
Na	0.38	0.32

(b)

Element	At. %	Wt.%
Al	73.79	76.34
C	18.57	8.55
Sn	0.83	3.76
Ag	0.82	3.39
Fe	0.72	1.55
Si	1.25	1.35
K	0.58	0.87
Ti	0.46	0.85
S	0.68	0.83
Mg	0.83	0.78
P	0.63	0.74
Ca	0.40	0.62
Na	0.44	0.37

FIGURE 11.13 SEM-EDS of (a) Al6063 (b) Al6063/10 wt.% PKS.

Source: Edoziuno, F. O., Adediran, A. A., Odoni, B. U., Utu, O. G., & Olayanju, A. (2021). Physico-chemical and morphological evaluation of palm kernel shell particulate reinforced aluminium matrix composites. *Materials Today: Proceedings*, *38*(2), 652–657. https://doi.org/10.1016/j.matpr.2020.03.641. Reproduced with permission.

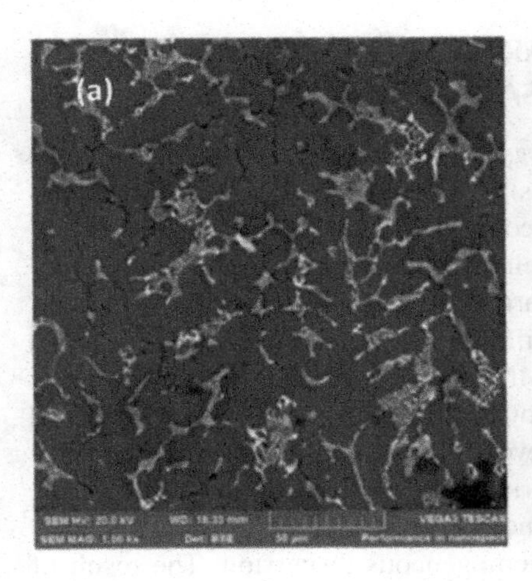

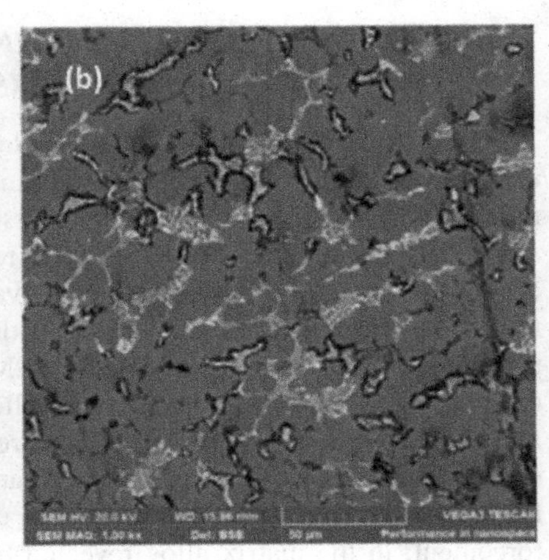

FIGURE 11.14 SEM micrographs of (a) A345 alloy (b) A345/4 wt.% PKSAnp.

Source: Aigbodion, V. S., & Ezema, I. C. (2020). Multifunctional A356 alloy/PKSAnp composites: Microstructure and mechanical properties. *Defence Technology*, *16*(3), 731–736. https://doi.org/10.1016/j.dt.2019.05.017. Reproduced with permission.

the matrix and reinforcements during double stir casting via the escape of trapped air bubbles in the slurry, thereby reducing porosity during processing. The elements detected through the EDS were traced to the agglomeration of Al6063, SiC, and PKSA chemical compositions. Aigbodion and Ezema (2020) examined the composites developed to contain two different phases of gray for the metal matrix and black for PKSA nanoparticles (PKSAnp) as reinforcement (Figure 11.14). The black phase was observed to be dispersed within the boundaries of the intermetallic phases of $Al_{12}Mg_7$ and Mg_2Si, as indicated by the EDS.

11.11 CORROSION OVERVIEW

Aqueous corrosion is a term required to explain the process by which materials or metals deteriorate when exposed to an aggressive environment. This process leads to surface damage in a hostile environment. Furthermore, corrosion is a chemical and/or electrochemical oxidation process where electrons are transferred to the other species in the system by the metal, thereby undergoing a change in valency from zero to more positive values. The electrolytes from these environments may be liquid, gas, or hybrid soil-liquid (Perez, 2004). An electrolyte contains charged ions, which are both positive (cations) and negative (anions), forming a conductive solution. An ion is said to be an atom that has gained or lost at least one outer electron and carries an electrical charge. Hence, at least two reactions (cathodic and anodic) are expected to occur in any particular corrosive environment in a corrosion process though without net current flow in the process. Oxidation and reduction reactions are expected to occur on the metal. When metal oxidation occurs, this is through the anodic reaction, and the cathodic reaction brings about reduction. The anodic reaction (oxidation) implies that electrons are lost, while the cathodic reaction (reduction) means electrons are gained or accepted.

11.11.1 Corrosion Classification

Corrosion can be classified into the following:

1. **General corrosion**: This occurs when the surface area of a metal or alloy entirely corrodes in an environment including liquid electrolyte (chemical solution, liquid metal), gaseous electrolyte (air, CO_2, etc.), or a hybrid electrolyte (solid and water, biological organisms, etc.). This categorization of general corrosion covers biological corrosion, high-temperature corrosion, liquid-metal corrosion, molten-salt corrosion, stray current corrosion, and galvanic corrosion (which happens between two incompatible metals or alloys) (Perez, 2004).
2. **Localized corrosion**: It is a form of corrosion that has been found to be more challenging to manage compared to general corrosion. This corrosion attacks specific sections of an exposed surface area in an appropriate electrolyte. Further classifications in this category are biological corrosion, selective leaching corrosion, pitting corrosion, crevice corrosion, and filiform corrosion (Perez, 2004).

In general, all the corrosion forms that engineering structures encounter have a common mechanism of oxidation usually denoted as an anodic reaction. Corrosion manifests its natural or forced behavior in different forms as stated above. Corrosion control is usually performed through numerous other processes and by coating to suppress the oxidation reaction.

11.11.2 Do Aluminum or AMCs Corrode?

Aluminum and its alloy are metals that have been known to have good corrosion resistance properties (Prasad and Asthana, 2004; Falcon et al., 2011; Ratna Kumar et al., 2017) in almost all corrosive environments owing to passivation. They easily form a protective film oxide to shield against corrosion. However, when AMCs are produced, there is a higher tendency to corrode more than the pure matrix. As reported by Cheng et al. (2007), MMCs (in which AMCs are inclusive) have less corrosion-resistant properties compared to conventional alloys because their corrosion behaviors are modified based on different factors. These factors include alloy composition, the microstructure of the matrix, the dispersoid and the matrix, matrix material distribution in the composite, the nature of the matrix-reinforcement interface, and the processing technique employed in the fabrication of the composite (Almomani et al., 2013; Bieniaś et al., 2003). As reported by Zakaria (2014), corrosion of composites is an essential parameter that determines its structural engineering applications in industry. With the addition of reinforcements, corrosion may be accelerated due to physical, chemical, or electrochemical interactions of the matrix with the reinforcement (Zakaria, 2014). It was further stated that the corrosion along the interface of a particle-matrix can result in rapid penetration of the matrix along the large interfacial regions. This will invariably result in improved corrosion of AMCs when compared with the monolithic alloys' corrosion. Corrosion is also accelerated through the galvanic interactions between the base metal and the reinforcing particulates. In addition, a crevice attack of the interface between the metal and reinforcement can result in structural and compositional heterogeneities within the matrix. Therefore, corrosion of AMCs, if not quickly attended to and controlled to a permissible limit, can lead to catastrophic failures and, hence, hamper their usage in a corrosive environment, mostly in stress-related areas.

AMC corrosion has been examined; however, published information on the corrosion of composites made of aluminum is frequently conflicting. The availability of various aluminum alloy matrixes and different combinations of reinforcement types may be partially responsible for this, which may result in the produced AMCs exhibiting different corrosion phenomena. Moreover, AMCs' processing parameters have a huge influence on composite microstructure and, consequently, on their corrosion performance. Corrosion studies have been carried out mostly on AMCs where synthetic reinforcement materials have been utilized (Alaneme and Bodunrin, 2011; Zakaria, 2014). Recent studies have centered on studying the corrosion behavior of AMCs with hybrid reinforcements of synthetic and ashes of agro-residues (Alaneme et al., 2015, 2018). However, the corrosion behavior of composites produced from hybrid reinforcements of SiC and PKSA has not been investigated. This review partly examines this area where the produced samples were subjected to aggressive environments.

In studying the corrosion performance of AMCs, the following methods are used in determining the corrosion rate, which are the weight change method (Gravimetric method) and electrochemical methods. These two methods have been used extensively to determine the corrosion rates of different aluminum alloys as well as AMCs (Zakaria, 2014). The gravimetric method is usually utilized in determining the weight loss of specimens in aggressive environments. The initial weight is measured before immersing the alloy/AMC sample in a corrosive environment. After a specified period, the sample is removed and cleaned following standard procedures and re-weighed to have the final weight. More so, the corrosion characteristics such as corrosion potential (E_{corr}), pitting potential (E_{pit}), and re-passivation potential (E_{rp}) of the samples can be determined from the electrochemical studies (Bieniaś et al., 2003). The corrosion characteristics of Mg-Al/TiC composites were carried out by Falcon et al. (2011) in which the experiments were performed in NaCl solution. For assessing corrosion behavior, electrochemical methods were used. The results from these techniques pointed to the fact that the composite samples showed lower corrosion rates than the base alloy. It was also discovered that there were galvanic effects, which accelerated the rate of corrosion between the matrix and the reinforcement used (TiC). More so, the base alloy was seen to have a high propensity to suffer from pitting corrosion compared to the composite. Ratna Kumar et al. (2017) investigated the corrosion tendency of AMC reinforced with multiwall carbon nanotube (MWCNT). AA5083 alloy was used as the matrix. The gravimetric method (weight change method) only was utilized for the corrosion study. Since AA5083/MWCNT can be used in a more corrosive environment, the immersion test was carried out using

90 mL HCl. It was discovered that the composite's corrosion resistance was higher than that of the base alloy. This was expected, as the major cathodic reaction in such a system was hydrogen evolution reaction, but oxygen reduction as the cathodic reaction may occur as well. Hydrogen evolution reaction normally takes place in two steps, as shown in Equations (11.1) and (11.2):

$$H^+ + e^- \rightarrow H \tag{11.1}$$

$$H + H \rightarrow H_2 \tag{11.2}$$

The hydrogen atom generated in the first step may react with aluminum to form the hydride, AlH_3, along with the well-known passivation oxide, Al_2O_3. This protective oxide layer serves as a physical barrier and indicates the thermodynamic stability of aluminum alloys in corrosive situations, which, however, when damaged, is capable of regenerating itself in oxidizing environments. Aluminum easily forms a strong hydride (Parshad, 1944; Perrault, 1979). This hydride is usually found sandwiched between the Al substrate and the protective passivation oxide layer, Al_2O_3, thus giving a second layer of protection on the substrate against the incursion of the aggressive environment. Composites are usually tailor-made for specific structural and other applications to obtain various relative advantages over other competing materials. Other researchers, such as Zakaria (2014), experimented with the corrosion of Al/SiC composite in 3.5% NaCl at various temperature regimes. The composite showed better corrosion resistance than the base alloy at ambient temperatures. However, reductions in reinforcement particle sizes, as well as reinforcement volume fraction increases, witnessed a decline in the corrosion resistance of the Al/SiC composite, even at elevated temperatures of 50°C and 70°C.

It was reported that Al/SiC inclusions are cathodic in most instances to Al alloys (Sukiman et al., 2012). Thus, the proliferation of these as cathodic reaction spots for the anodic dissolution of the Al matrix will constitute more electron demand on the anodic matrix. The more the cathodic spots demand electrons, the higher the corrosion rate of the composite. Also, it is generally known that for every 10°C rise in temperature for a chemical or electrochemical reaction, there must be a corresponding 50%–100% increase in reaction rates. Thus, the findings of the researchers buttress the fact that composites are engineered for specific environments, and apart from their physico-mechanical/chemical attributes, electrochemical corrosion/mitigation preparedness must be put in perspective. It is good to find ways by which AMC corrosion should be mitigated. Hence, the next section is dedicated to the mitigation methods against AMC corrosion.

11.11.3 Mitigation Against AMC Corrosion

The reduction, by suitable approaches, of the chemical or electrochemical disintegration of advanced engineering materials such as AMCs is key to their deployment. This is due to the devastating effects that corrosion has on metals and their alloys. Therefore, material degradation due to chemical and/or electrochemical reactions is required to be tackled and mitigated. Oki (2015) and Hihara (2010) stated some of the ways to alleviate corrosion, which are through the utilization of paints, corrosion inhibitors, cathodic protection, engineering coatings (such as chromate conversion coating, anodization, and so on). Of interest among all these protection techniques is the conversion coating.

Conversion coatings are finishes used by the metal industry to coat metals and increase the substrate's ability to resist corrosion. Additionally, they aid in enhancing the adhesion properties of subsequently applied organic coatings on treated surfaces. They are coatings for metals in which the coating solution subjects the metals' surfaces to a chemical or electrochemical reaction, transforming them into a decorative and/or protective layer. Examples of conversion coatings include chromate and phosphate conversion coatings, bluing, parkerizing, black oxide coatings on steel, and anodizing. These serve as protection against corrosion, as paint primers, and add decorative color. To conversion-coat, the metallic material can be immersed into a chemical solution with or without the application of electric current within the coating system. As a result, there will be growth of a coating on the surface of the substrate owing to the reaction between the metal/alloy and the chemical, which may be assisted by the passage of current as with anodizing. The coating is an integral part of the component surface and not just deposited over the surface. Conversion coatings can be classified into two. The chemical conversion coatings consist of phosphate coatings formed on steel or zinc, chromate coatings formed on aluminum, cadmium, zinc, copper, magnesium, and silver, and oxide coatings formed on copper, steel, iron, and zinc alloys. The second classification is electrochemical conversion coatings (anodization). This anodization helps to convert the surface of certain aluminum, magnesium, titanium, and zinc alloys to metal oxide. This method is characterized by the immersion of the material in an acidic solution and passing electricity through the material, where the anode in the electrical circuit is the metal/alloy of interest.

A major method in protecting Al alloys and AMCS against localized and galvanic corrosion is through treatment with chemical conversion coatings using a solution containing chromate ions. According to Oki (2015), there is ease of application giving excellent service performance when chromate conversion coating is utilized. This has given this method high popularity in the high-safety sectors of the metal finishing industry, including roofing and aviation industries. The coatings of hexavalent chromium are versatile, as they have good corrosion protection and are cost-effective. This outstanding chromate's resistance to corrosion has been consistent over very broad electrolyte concentrations and pH ranges. Alternative inhibitory substances have not yet been able to match this characteristic. Chromium can restrict metal dissolution rate because they are both anodic and cathodic inhibitors. This also helps to reduce the reduction reaction rate (oxygen and water reduction) in many environments. It is said to be a potent corrosion inhibitor, as it has been incorporated into corrosion protection primers and coatings. However, due to the high toxicity and carcinogenic nature of chromium/chromate ions, numerous novel conversion coating methods have been developed as a result of research into non-toxic corrosion inhibitors (Oki, 2015; Gharbi et al., 2018). In order to mitigate against this carcinogenic effect of chromium/chromate ions, alternatives to chromium technology for conversion coatings are being developed, though most are not as effective as chromate conversion coatings. Some of these alternatives are rare-earth-metal-based coatings, including cerium (-nitrate, -sulfate, -chloride), organic coatings, lithium-based coatings, vanadate-based coatings, and nanocomposites, such as hybrid sol-gel, organic inhibitors, and polymer coatings; phosphate coatings, such as strontium phosphate (SrP); and metal-rich primers (Gharbi et al., 2018).

Recently, the corrosion behavior of MMCs reinforced with PKSA and SiC in both acidic and salt environments was evaluated by Ikubanni et al. (2022b, 2022c). Pitting corrosion was observed to be a common occurrence on the composites. The composites corroded faster in the H_2SO_4 solution compared to the NaCl solution. Based on these facts, in other to mitigate the corrosion effects on AMCs, many recent studies are ongoing to secure alternative chromate-free and non-toxic conversion coating on AMCs to mitigate corrosion activities. Potassium permanganate ($KMnO_4$) conversion coating was compared with chromate conversion coating. The former conversion coating performed fairly well in protecting the surface of the developed composites in the atmosphere and salt environment. Also, it enhanced the adhesion of paints on the conversion-coated surfaces without delamination.

11.12 FUTURE WORKS

The utilization of developed advanced engineering materials, including MMCs, is critical to materials scientists and engineers, and to the people in various industries (the end-users). Numerous challenges abound and must be overcome to achieve the goal of proper utilization of the synthesized MMCs. More studies are required in the areas of utilization of industrial and agro-waste as reinforcements in producing low-cost but high-quality MMCs. Various combinations of different agro-wastes, as well as industrial wastes, should be investigated using different processing routes. The ductility of MMCs is usually low; hence, ways to improve this property to enhance damage tolerance should be embarked on. MMCs have found utilization in aerospace and aviation industries, automobile industries, marine industries, and so on; however, limited utilizations are obtainable in locomotive industries. For new market opportunities, there should be growing use of MMCs in locomotive industries, as well as their further adoption in electric vehicle production. Further works are recommended in mitigating corrosion of MMCs to make them very suitable in areas that are not corrosion-friendly. New methods of processing MMCs, such as rheocasting, double layer feeding-stir-casting method, and so on, should be properly explored to harness their benefits and demerits.

11.13 CONCLUSION

1. This study has given an overview of different methods of synthesizing MMCs with different hybrid reinforcements and determination of the properties such as physical, mechanical, and tribological properties.
2. The corrosion overview of MMCs was highlighted, and PKSA reinforcement was specifically discussed, in which corrosion initiation begins from the secondary materials embedded in the MMCs.
3. Areas of utilization of MMCs were highlighted. With numerous available materials for reinforcing metal matrixes, there are hosts of new advanced engineering materials that could be developed by materials and design engineers using different reinforcement combinations to arrive at a predetermined and suitable property. This implied that the development of MMCs for various applications in different industries and commercialization has a great future.
4. Future areas of research and development on MMCs were discussed to include the utilization in locomotive industries and electric vehicle development. Lastly, various cost-reduction approaches are being developed by different researchers to lower the replacement cost of steel materials with Al MMCs.

REFERENCES

Adebayo, O. (2012). Asessment of palm kernel shells as aggregate in concrete and laterite blocks. *Journal of Engineering Studies and Research*, *18*(2), 88–93.

Adediran, A. A., Edoziuno, F. O., Adesina, O. S., Sodeinde, K. O., Ogunkola, A. B., Oyinloye, G. A., Nwaeju, C. C., & Akinlabi, E. T. (2022). Mechanical characterization and numerical optimization of aluminum matrix hybrid composite. *Materials Science Forum*, *1065*, 47–57. https://doi.org/10.4028/p-m21wne

Ahamed, A. A., Ahmed, R., Hossain, M. B., & Billah, M. (2016). Fabrication and characterization of aluminium-rice husk ash composite prepared by stir casting method. *Rajshahi University Journal of Science and Engineering*, *44*, 9–18. https://doi.org/10.3329/rujse.v44i0.30361

Aigbodion, V. S., & Ezema, I. C. (2020). Multifunctional A356 alloy/PKSAnp composites: Microstructure and mechanical properties. *Defence Technology*, *16*(3), 731–736. https://doi.org/10.1016/j.dt.2019.05.017

Akinlabi, E. T., Fono-Tamo, R. S., & Tien-Chien, J. (2019). Microstructural and dry sliding friction studies of Aluminium matrix composites reinforced PKS ash developed via friction stir processing. In C. Chesonis (Ed.), *The Minerals, Metals & Materials Series* (Light Meta, pp. 401–406). https://doi.org/10.1007/978-3-030-05864-7_51

Alaneme, K. K., Adegun, M. H., Archibong, A. G., & Okotete, E. A. (2019). Mechanical and wear behaviour of aluminium hybrid composites reinforced with varied aggregates of alumina and quarry dust. *Journal of Chemical Technology and Metallurgy*, *54*(6), 1361–1370.

Alaneme, K. K., Ademilua, B. O., & Bodunrin, M. O. (2013). Mechanical properties and corrosion behaviour of aluminium hybrid composites reinforced with silicon carbide and bamboo leaf ash tribology in Industry. *Tribology in Industry*, *35*(1), 25–35.

Alaneme, K. K., & Adewale, T. M. (2013). Influence of rice husk ash/silicon carbide weight ratios on the mechanical behaviour of aluminium matrix hybrid composites. *Tribology in Industry*, *35*(2), 163–172.

Alaneme, K. K., Adewale, T. M., & Olubambi, P. A. (2014). Corrosion and wear behaviour of Al – Mg – Si alloy matrix hybrid composites reinforced with rice husk ash and silicon carbide. *Journal of Materials Research and Technology*, *3*(1), 9–16. https://doi.org/10.1016/j.jmrt.2013.10.008

Alaneme, K. K., & Aluko, A. O. (2012). Fracture toughness (K_1C) and tensile properties of as-cast and age-hardened aluminium (6063)-silicon carbide particulate composites. *Scientia Iranica A*, *19*(4), 992–996. https://doi.org/10.1016/j.scient.2012.06.001

Alaneme, K. K., & Bodunrin, M. O. (2011). Corrosion behaviour of Alumina Reinforced Aluminium (6063) metal matrix composites. *Journal of Minerals and Materials Characterization and Engineering*, *10*(12), 1153–1165.

Alaneme, K. K., & Bodunrin, M. O. (2013). Mechanical behaviour of alumina reinforced AA 6063 metal matrix composites developed by two-stir casting process. *Acta Technic Corvininesis-Bulletin of Engineering*, *6*(3), 105–110.

Alaneme, K. K., Bodunrin, M. O., & Awe, A. A. (2018). Microstructure, mechanical and fracture properties of groundnut shell ash and silicon carbide dispersion strengthened aluminium matrix composites. *Journal of King Saud University - Engineering Sciences*, *30*(1), 96–103. https://doi.org/10.1016/j.jksues.2016.01.001

Alaneme, K. K., Eze, H. I., & Bodunrin, M. O. (2015). Corrosion behaviour of groundnut shell ash and silicon carbide hybrid reinforced Al-Mg-Si alloy matrix composites in 3.5 % NaCl and 0.3M H_2SO_4 solutions. *Leonardo Electronic Journal of Practices and Technologies*, *26*, 129–146.

Alaneme, K. K., & Olubambi, P. (2013). Corrosion and wear behaviour of rice husk ashdalumina reinforced Al-Mg-Si alloy matrix hybrid composites. *Journal of Materials Research and Technology*, *2*(2), 188–194.

Alaneme, K. K., & Sanusi, O. K. (2015). Microstructural characteristics, mechanical and wear behaviour of aluminium matrix hybrid composites reinforced with alumina, rice husk ash and graphite. *Engineering Science and Technology, an International Journal*, *18*(3), 416–422. https://doi.org/10.1016/j.jestch.2015.02.003

Allison, J. E., & Cole, G. S. (1993). Metal-matrix composites in the automotive industry: Opportunities and challenges. *Journal of Metals*, *45*, 19–24.

Almomani, M., Hassan, A. M., Qasim, T., & Ghaithan, A. (2013). Effect of process parameters on corrosion rate of friction stir welded aluminium SiC-Gr hybrid composites. *Corrosion Engineering Science and Technology*, *48*(5), 346–353. https://doi.org/10.1179/1743278213Y.0000000083

Ananda Murthy, H. C., & Singh, S. K. (2015). Influence of TiC particulate reinforcement on the corrosion behaviour of Al6061 metal matrix composites. *Advanced Materials Letters*, *6*(7), 633–640. https://doi.org/10.5185/amlett.2015.5654

Ansari, S. A., & Cho, M. H. (2016). Highly visible light responsive, narrow band gap TiO_2 nanoparticles modified by elemental red phosphorus for photocatalysis and photoelectrochemical applications. *Scientific Reports*, *6*(January), 1–10. https://doi.org/10.1038/srep25405

Anyaoha, K. E., Sakrabani, R., Patchigolla, K., & Mouazen, A. M. (2018). Critical evaluation of oil palm fresh fruit bunch solid wastes as soil amendments: Prospects and challenges. *Resources, Conservation and Recycling*, *136*(September), 399–409. https://doi.org/10.1016/j.resconrec.2018.04.022

Anyika, C., Asilayana, N., Asri, M., Abdul, Z., Adibah, Y., & Jafariah, J. (2017). Synthesis and characterization of magnetic activated carbon developed from palm kernel shells. *Nanotechnology for Environmental Engineering*, *16*(2), 1–25. https://doi.org/10.1007/s41204-017-0027-6

Aweda, J. O., & Adeyemi, M. B. (2009). Determination of temperature distribution in squeeze cast aluminium using the semi-empirical equations' method. *Journal of Materials Processing Technology*, *209*, 5751–5759.

Baradeswaran, A., & Elaya Perumal, A. (2014). Study on mechanical and wear properties of Al 7075/Al2O3/graphite hybrid composites. *Composites Part B: Engineering*, *56*, 464–471.

Bhoi, N. K., Singh, H., & Pratap, S. (2020). Developments in the aluminum metal matrix composites reinforced by micro/nanoparticles – A review. *Journal of Composite Materials*, *54*(6), 813–833. https://doi.org/10.1177/0031998319865307

Bieniaś, J., Walczak, M., Surowska, B., & Sobczak, J. (2003). Microstructure and corrosion behaviour of Aluminium fly ash composites. *Journal of Optoelectronics and Advanced Materials*, *5*(2), 493–502.

Bodunrin, M.O., Alaneme, K.K., Chown, L. H. (2015). Aluminium matrix hybrid composites: a review of reinforcement philosophies; mechanical, corrosion and tribological characteristics. *Journal of Materials Research and Technology*, *4*(4), 434–445. https://doi.org/10.1016/j.jmrt.2015.05.003

Boopathi, M., Arulshri, K. P., & Iyandurai, N. (2013). Evaluation of Mechanical Properties of Aluminium Alloy2024 Reinforced with Silicon Carbide and Fly Ash Hybrid Metal Matrix Composites. *American Journal of Applied Sciences*, *10*(3), 219–229. https://doi.org/10.3844/ajassp.2013.219.229

Casati, R., & Vedani, M. (2014). Metal matrix composites reinforced by nano-particles-A review. *Metals*, *4*(1), 65–83.

Chawla, N., Shen, Y. (2001). Mechanical behaviour of particle reinforced metal matrix composites. *Advanced Engineering Materials*, *3*(6), 357–370.

Cheng, Y. L., Chen, Z. H., Wu, H. L., & Wang, H. M. (2007). The corrosion behaviour of the Aluminium alloy 7075/SiCp metal matrix composite prepared by spray deposition. *Materials and Corrosion*, *58*(4), 280–284. https://doi.org/10.1002/maco.200604003

Das, D. K., Mishra, P. C., Singh, S., & Pattanaik, S. (2014). Fabrication and heat treatment of ceramic-reinforced aluminium matrix composites-A review. *International Journal of Mechanical and Material s Engineering*, *9*(1), 1–15.

Davis, J. R. (2001). Light metals and alloys-aluminium and aluminium alloys. *Alloying: Understanding the Basics*, 351–416. https://doi.org/10.1361/autb2001p351

Demirbas, A. (2004). Combustion characteristics of different biomass fuels. *Progress in Energy and Combustion Science*, *30*(2), 219–230. https://doi.org/10.1016/j.pecs.2003.10.004

Deravaju, A., Kumar, A., & Kotiveerachari, B. (2013). Influence of addition of Grp/Al2O3p with SiCp on wear properties of aluminium alloy 6061-T6 hybrid composites via friction stir processing. *Transactions of Nonferrous Metals Society of China2*, *23*(5), 1275–1280.

Dhanashekar, M., & Senthil Kumar, V. S. (2014). Squeeze casting of aluminium metal matrix composites-An overview. *12th Global Congress on Manufacturing and Management, Procedia Engineering*, *97*, 412–420.

Dinaharan, I., Nelson, R., Vijay, S.J., & Akinlabi, E.T. (2016). Microstructure and wear characterization of Aluminium matrix composites reinforced with industrial waste fly ash particulates synthesized by friction stir processing. *Materials Characterization*, *118*, 149–158. https://doi.org/10.1016/j.matchar.2016.05.017

Dinaharan, I., Kalaiselvan, K., & Murugan, N. (2017). Influence of rice husk ash particles on microstructure and tensile behaviour of AA6061 Aluminium matrix composites produced using friction stir processing. *Composites Communications*, *3*, 42–46. https://doi.org/10.1016/j.coco.2017.02.001

Dixit, P., & Suhane, A. (2022). Aluminum metal matrix composites reinforced with rice husk ash: A review. *Materials Today: Proceedings*, *62*(6), 4194–4201. https://doi.org/10.1016/j.matpr.2022.04.711

Edoziuno, F. O., Adediran, A. A., Odoni, B. U., Utu, O. G., & Olayanju, A. (2021). Physico-chemical and morphological evaluation of palm kernel shell particulate reinforced aluminium matrix composites. *Materials Today: Proceedings*, *38*(2), 652–657. https://doi.org/10.1016/j.matpr.2020.03.641

Escalera-Lozano, R., Gutierrez, C., Pech-Canul, M. A., & Pech-Canul, M. I. (2008). Degrada- tion of Al/SiCp composites produced with rice-hull ash and aluminium cans. *Waste Management*, *28*, 389–395.

Escalera-Lozano, R., Gutiérrez, C. A., Pech-Canul, M. A., & Pech-Canul, M. I. (2007). Corrosion characteristics of hybrid Al/SiCp/MgAl2O4 composites fabricated with fly ash and recycled Aluminium. *Materials Characterization*, *58*, 953–960. https://doi.org/10.1016/j.matchar.2006.09.012

Falcon, L. A., Bedolla B. E., Lemus, J., Leon, C., Rosales, I., & Gonzalez-Rodriguez, J. G. (2011). Corrosion behaviour of Mg-Al/TiC composites in NaCl solution. *International Journal of Corrosion*, *2011*. https://doi.org/10.1155/2011/896845

Fatchurrohman, N., Sulaiman, S., Sapuan, S. M., Ariffin, M. K. A., & Baharuddin, B. T. H. T. (2015). Analysis of a metal matrix composites automotive component. *International Journal of Automotive and Mechanical Engineering*, *11*, 2531–2540. doi.org/10.15282/ijame.11.2015.32.0213

Fatile, O. B., Akinruli, J. I., & Amori, A. A. (2014). Microstructure and mechanical behaviour of Stir-Cast Al-Mg-Si alloy matrix hybrid composite reinforced with Corn Cob Ash and Silicon Carbide. *International Journal of Engineering and Technology Innovation*, *4*(4), 251–259.

Fono-Tamo, R. S., Idowu, O. O., & Koya, F. O. (2014). Development of pulverized palm kernel shells based particleboard. *International Journal of Material and Mechanical Engineering*, *3*(3), 54. https://doi.org/10.14355/ijmme.2014.0303.01

Fono-Tamo, R. S., & Koya, O. A. (2013). Characterisation of pulverised palm kernel shell for sustainable waste diversification. *International Journal of Scientific & Engineering Research*, *4*(4), 2229–5518. https://doi.org/2229-5518

García, J. R., Sedran, U., Zaini, M. A. A., & Zakaria, Z. A. (2018). Preparation, characterization, and dye removal study of activated carbon prepared from Palm Kernel Shell. *Environmental Science and Pollution Research*, *25*(6), 5076–5085. https://doi.org/10.1007/s11356-017-8975-8

Gharbi, O., Thomas, S., Smith, C., & Birbilis, N. (2018). Chromate replacement: what does the future hold? *Npj Materials Degradation*, *2*(1), 23–25. https://doi.org/10.1038/s41529-018-0034-5

Gillani, F., Khan, M. Z., & Shah, O. R. (2022). Sensitivity analysis of reinforced aluminum based metal matrix composites. *Materials*, *15*(12), 4225. https://doi.org/10.3390/ma15124225

Haider, K., Alam, A., Redhewal, A., & Saxena, V. (2015). Investigation of mechanical properties of aluminium based metal matrix investigation of mechanical properties of aluminium based metal matrix composites reinforced with Sic & Al_2O_3. *International Journal of Engineering Research and Applications*, *5*(9), 63–39.

Haque, S., Bharti, P. K., & Ansari, A. H. (2014). Mechanical and Machining Properties Analysis of Al6061-Cu-Reinforced SiC_P; Metal Matrix Composite. *Journal of Minerals and Materials Characterization and Engineering*, *2*(1), 54–60. https://doi.org/10.4236/jmmce.2014.21009

Hardjasaputra, H., Fernando, I., Indrajaya, J., Cornelia, M., & Rachmansyah. (2018). The effect of using palm kernel shell ash and rice husk ash on geopolymer concrete. *MATEC Web of Conferences*, *251*. https://doi.org/10.1051/matecconf/201825101044

Hihara, L. H. (2010). Corrosion of metal matrix composites. *Shreir's Corrosion*, *6*, 2250–2269. https://doi.org/10.1016/B978-044452787-5.00110-4

Hwu, B. K., Lin, S. J., & Jahn, M. T. (1996). Effects of process parameters on the properties of squeeze-cast SiC-6061 Al metal-matrix composite. *Materials Science and Engineering, A207*, 135–141.

Ibhadode, A. O. A., & Dagwa, I. M. (2008). Development of asbestos-free friction lining material from palm kernel shell. *Journal of the Brazilian Society of Mechanical Sciences and Engineering, 30*(2), 166–173. https://doi.org/10.1590/S1678-58782008000200010

Ibikunle, R. A., Ikubanni, P. P., Agboola, O. O., & Ogunsemi, B. T. (2018). Development and performance evaluation of palm nut cracker. *Leonardo Electronic Journal of Practices and Technologies, 33*, 219–234.

Ikele, U. S., Alaneme, K. K., & Akinlabi, O. (2022). Mechanical behaviour of stir cast aluminium matrix composites reinforced with silicon carbide and palmkernel shell ash. *Manufacturing Review, 9*(12), 1–13. https://doi.org/10.1051/m/review/2022011

Ikubanni, P., Oki, M., Adeleke, A., Adesina, O., & Omoniyi, P. (2021a). Physico-tribological characteristics and wear mechanism of hybrid reinforced a16063 matrix composites. *Acta Metallurgica Slovaca, 27*(4), 172–179. https://doi.org/10.36547/ams.27.4.1084

Ikubanni, P., Oki, M., Adeleke, A., Adesina, O., Omoniyi, P., & Akinlabi, E. (2022a). Electrochemical studies of the corrosion behaviour of Al/SiC/PKSA hybrid composites in 3.5% NaCl solution. *Journal of Composites Science, 6*(10), 1–13. https://doi.org/10.3390/jcs6100286

Ikubanni, P., Oki, M., Adeleke, A., Ajisegiri, E., & Fajobi, M. (2022b). Corrosion behaviour of Al/SiC/PKSA hybrid composites in 1.0 H_2SO_4 environment using potentiodynamic polarization technique. *Acta Metallurgica Slovaca, 28*(4), 169–171. https://doi.org/10.36547/ams.28.4.1561

Ikubanni, P., Oki, M., Adeleke, A., Omoniyi, P., Ajisegiri, E., & Akinlabi, E. (2022c). Physico-mechanical properties and microstructure responses of hybrid reinforced Al6063 composites to PKSA/SiC inclusion. *Acta Metallurgica Slovaca, 28*(1), 25–32. https://doi.org/10.36547/ams.28.1.1340

Ikubanni, P. P., Oki, M., & Adeleke, A. A. (2020a). A review of ceramic/bio-based hybrid reinforced aluminium matrix composites. *Cogent Engineering, 7*(1), 1–19. https://doi.org/10.1080/23311916.2020.1727167

Ikubanni, P. P., Oki, M., Adeleke, A. A., Adediran, A. A., & Adesina, O. S. (2020b). Influence of temperature on the chemical compositions and microstructural changes of ash formed from palm kernel shell. *Results in Engineering, 8*. https://doi.org/10.1016/j.rineng.2020.100173

Ikubanni, P. P., Oki, M., Adeleke, A. A., Adediran, A. A., Agboola, O. O., Babayeju, O., Egbo, N., & Omiogbemi, I. M. B. (2021b). Tribological and physical properties of hybrid reinforced aluminium matrix composites. *Materials Today: Proceedings, xxxx*. https://doi.org/10.1016/j.matpr.2021.03.537

Ikubanni, P. P., Oki, M., Adeleke, A. A., & Agboola, O. O. (2021c). Optimization of the tribological properties of hybrid reinforced aluminium matrix composites using Taguchi and Grey's relational analysis. *Scientific African*. https://doi.org/10.1016/j.sciaf.2021.e00839

Ikubanni, P. P., Oki, M., Adeleke, A. A., & Omoniyi, P. O. (2021d). Synthesis, physico-mechanical and microstructural characterization of Al6063/SiC/PKSA hybrid reinforced composites. *Scientific Reports, 11*. https://doi.org/10.1038/s41598-021-94420-0

Ikumapayi, O. M., & Akinlabi, E. T. (2018). Composition, characteristics and socio-economic benefits of palm kernel shell exploitation-An overview. *Journal of Environmental Science and Technology, 11*(5), 220–232. https://doi.org/10.3923/jest.2018.220.232

Imoisili, P. E., Ukoba, K. O., & Jen, T. (2020). Synthesis and characterization of amorphous mesoporous silica from palm kernel shell ash. *Boletín de La Sociedad Española de Cerámica y Vidrio*, *59*(4), 159–164. https://doi.org/10.1016/j.bsecv.2019.09.006

Ishola, M., Oladimeji, O., & Paul, K. (2017). Development of ecofriendly automobile brake pad using different grade sizes of palm kernel shell powder. *Current Journal of Applied Science and Technology*, *23*(2), 1–14. https://doi.org/10.9734/cjast/2017/35766

Jeykrishnan, J., Ramnath, B. V., Savariraj, X. H., Prakash, R. D., Rajan, V. R. D., & Kumar, D. D. (2016). Investigation on tensile and impact behaviour of aluminium base silicon carbide metal matrix composites. *Indian Journal of Science and Technology*, *9*(37), 2–5. https://doi.org/10.17485/ijst/2016/v9i37/101979

Kala, H., Mer, K. K. S., & Kumar, S. (2014). A review on mechanical and tribological behaviours of stir cast aluminium matrix composites. *Procedia Materials Science*, *6*, 1951–1960. https://doi.org/10.1016/j.mspro.2014.07.229

Kanth, U. R., Rao, P. S., & Krishna, M. G. (2019). Mechanical behaviour of flyash/SiC particles reinforced Al-Zn alloy-based metal matrix composites fabricated by stir casting method. *Journal of Materials Research and Technology*, *8*(1), 737–744. https://doi.org/10.1016/j.jmrt.2018.06.003

Kareem, A., Qudeiri, J. A., Abdudeen, A., Ahammed, T., & Ziout, A. (2021). A review on AA6061 metal matrix composites produced by stir casting. *Materials*, *14*(1), 175. https://doi.org/10.3390/ma14010175

Kopeliovich, D. (2012a). Liquid state fabrication of metal matrix composites. www.substech.com/dokuwiki/doku.php?id=liquid_state_fabrication_of_metal_matrix_composites (Accessed January 12 2023).

Kopeliovich, D. (2012b). In-situ fabrication of metal matrix composites. www.substech.com/dokuwiki/doku.php?id=in_situ_fabrication_of_metal_matrix_composites (Accessed January 12 2023).

Kopeliovich, D. (2012c). Solid state fabrication of metal matrix composites. www.substech.com/dokuwiki/doku.php?id=solid_state_fabrication_of_metal_matrix_composites (Accessed January 12 2023).

Krishna, M. V., & Xavior, A. M. (2014). An Investigation on the Mechanical Properties of Hybrid Metal Matrix Composites. *Procedia Engineering*, *97*, 918–924. https://doi.org/10.1016/j.proeng.2014.12.367

Krishnappa, G. B., Arunkumar, K. N., & Pasha, M. S. (2018). A study on hardness and wear behaviour of untreated and cryogenically treated Al-SiC and Al-Gr metal matrix composite. *MATEC Web of Conferences*, *144*. https://doi.org/10.1051/matecconf/201714402020

Krupakara, P. V., & Ravikumar, H. R. (2015). Corrosion Characterization of Aluminium 6061/Red Mud Metal Matrix Composites in Sea Water. *International Journal of Advanced Research in Chemical Science*, *2*(6), 52–55.

Kumar, B. P., & Birru, A. K. (2018). Tribological behavior of aluminium metal matrix composite with addition of bamboo leaf ash by GRA-Taguchi method. *Tribology in Industry*, *40*(2), 311–325. https://doi.org/10.24874/ti.2018.40.02.14

Kumar, G. S. P., Keshavamurthy, R., Kumari, P., & Dubey, C. (2016). Corrosion behaviour of TiB_2 reinforced aluminium based in situ metal matrix composites. *Perspectives in Science*, *8*, 172–175. https://doi.org/10.1016/j.pisc.2016.04.025

Kurumlu, D., Payton, E. J., Young, M. L., Schöbel, M., Requena, G., & Eggeler, G. (2012). High-temperature strength and damage evolution in short fiber reinforced Aluminium alloys studied by miniature creep testing and synchrotron microtomography. *Acta Materialia*, *60*(1), 67–78.

Lancaster, L., Lung, M. H., & Sujan, D. (2013). Utilization of Agro-industrial waste metal matrix composites: Towards sustainability. *World Academy of Science, Engineering and Technology*, *73*, 1136–1144.

Li, B.T., Lavernia, E.J., Lin, Y., Chen, F., and Zhang, L. (2016). Spray forming of MMCs. In: Reference Module in Materials Science and Materials Engineering, p.6. https://doi.org/10.1016/B978-0-12-803581-8.03884-4

Loto, R. T., & Babalola, P. (2017). Corrosion polarization behaviour and microstructural analysis of AA1070 aluminium silicon carbide matrix composites in acid chloride concentrations. *Cogent Engineering*, *4*(1), 1–11. https://doi.org/10.1080/23311916.2017.1422229

Macke, J., Schultz, B., & Rohatgi, P. (2012). Metal matrix composites offer the automotive industry an opportunity to reduce vehicle weight, improve performance. *Advanced Materials and Processes*, *170*(3), 19–23.

Mahendra, K. V., & Radha Krishna, K. (2010). Characteristics of stir cast Al.Cu (fly ash +SiC) hybrid metal matrix composites. *Journal of Composite Materials*, *44*(8), 989–1005. https://doi.org/10.1177/0021998309346386

Matthews, F. L., Davies, G. A., Hitchings, D., & Soutis, C. (2000). *Finite Element Modelling of Composite Materials and Structures*. Elsevier.

Mavhungu, S. T., Akinlabi, E. T., Onitiri, M. A., & Varachia, F. M. (2017). Aluminium Matrix Composites for Industrial Use: Advances and Trends. *International Conference on Sustainable Materials Processing and Manufacturing, SMPM 2017, 23-35 January 2017, Kruger National Park Published in Procedia Manufacturing*, *7*, 178–182. https://doi.org/10.1016/j.promfg.2016.12.045

Mgbemena, C. O., Mgbemena, C. E., & Okwu, M. O. (2014). Thermal stability of pulverized palm kernel shell (PKS) based friction lining material locally developed from spent waste. *ChemXpress*, *5*(4), 115–122.

Misnon, I. I., Zain, N. K. M., & Jose, R. (2019). Conversion of oil palm kernel shell biomass to activated carbon for supercapacitor electrode application. *Waste and Biomass Valorization*, *10*(6), 1731–1740. https://doi.org/10.1007/s12649-018-0196-y

Mohamad, M., Ramli, A., Misi, S., & Yusup, S. (2011). Steam gasification of Palm Kernel Shell (PKS): Effect of Fe/BEA and Ni/BEA catalysts and steam to biomass ratio on composition of gaseous products. *International Journal of Chemical, Molecular, Nuclear, Materials and Metallurgical Engineering*, *5*(12), 1085–1090. http://www.waset.org/publications/12818

Moona, G. C., Walia, R. S., Vikas, R., & Sharma, R. (2018). Aluminium metal matrix composites: A retrospective investigation. *International Journal of Pure and Applied Physics*, *56*, 164–175.

Mortimer, R., Saj, S., & David, C. (2018). Supporting and regulating ecosystem services in cacao agroforestry systems. *Agroforestry Systems*, *92*(6), 1639–1657. https://doi.org/10.1007/s10457-017-0113-6

Mussatto, A., Ahad, I. U. I., Mousavian, R. T., Delaure, Y., & Brabazon, D. (2021). Advanced production routes for metal matrix composites. *Engineering Reports*, *3*(5), e12330. https://doi.org/10.1002/eng2.12330

Narayan, S., & Rajeshkannan, A. (2017). Hardness, tensile and impact behaviour of hot forged aluminium metal matrix composites. *Journal of Materials Research and Technology*, *6*(3), 213–219. https://doi.org/10.1016/j.jmrt.2016.09.006

Narula, C. K., Allison, J. E., Bauer, D., & Gandhi, H. S. (1996). Materials chemistry issues related to advanced materials applications in the automotive industry. *Chemistry of Materials*, *26*, 48.

Ninduangdee, P., & Kuprianov, V. I. (2014). Fluidized-bed combustion of biomass with elevated alkali content: A comparative study between two alternative bed materials. *International Journal of Energy and Power Engineering*, *8*(4), 267–274.

Nturanabo, F., Masu, L., & Baptist Kirabira, J. (2019). Novel applications of aluminium metal matrix composites. *Aluminium Alloys and Composites [Working Title]*. https://doi.org/10.5772/intechopen.86225

Obi, O. F. (2015). Evaluation of the physical properties of composite briquette of sawdust and palm kernel shell. *Biomass Conversion and Biorefinery*, *5*(3), 271–277. https://doi.org/10.1007/s13399-014-0141-7

Oki, M. (2015). Performance of Tannin/Glycerol-Chromate hybrid conversion coating on aluminium. *Journal of Materials Science and Chemical Engineering*, *3*(7), 1–6. https://doi.org/10.4236/msce.2015.37001

Oki, M., Adediran, A.A., Olayinka, S., & Ogunsola, O. (2017). Development and performance of hybrid coatings on aluminium alloy. *Journal of Electrochemical Science and Engineering*, *7*(3), 131–138. https://doi.org/10.5599/jese.347

Okoroigwe, E. C., Saffron, C. M., & Kamdem, P. D. (2014). Characterization of palm kernel shell for materials reinforcement and water treatment. *Journal of Chemical Engineering and Material Science*, *5*(1), 1–6. https://doi.org/10.5897/JCEMS2014.0172

Oladele, I. O., & Okoro, A. M. (2016). The effect of palm kernel shell ash on the mechanical properties of as-cast aluminium alloy matrix composites. *Leonardo Journal of Sciences*, *28*, 15–30.

Olutoge, F. A., Quadri, H.A., & Olafusi, O. S. (2012). Investigation of the strength properties of palm kernel shell ash concrete. *Engineering Technology & Applied Science Research*, *2*(6), 315–319.

Osei, D. Y. (2013). Pozzolana and palm kernel shells as replacements of portland cement and crushed granite in concrete. *International Journal of Engineering Inventions*, *2*(10), 1–5.

Oti, J. E., Kinuthia, J. M., Robinson, R., & Davies, P. (2015). The use of palm kernel shell and ash for concrete production. *International Science Index, Civil and Environmental Engineering*, *9*(1), 263–270. https://waset.org/publications/10000699/the-use-of-palm-kernel-shell-and-ash-for-concrete-production

Padmavathi, K. R., & Ramakrishnan, R. (2014). Tribological behaviour of aluminium hybrid metal matrix composite. In *Procedia Engineering* (Vol. 97, pp. 660–667). https://doi.org/10.1016/j.proeng.2014.12.295

Parshad, R. (1944). Formation of aluminium hydride layers on aluminium. *Nature*, *154*, 178. https://doi.org/doi.org/10.1038/154178a0

Parswajinan, C., Ramnath, B.V., Abishek, S., Niharishsagar, B., & Sridhar, G. (2018a). Hardness and impact behaviour of aluminium metal matrix composite Hardness and impact behaviour of aluminium metal matrix composite. *IOP Conf. Series: Materials Science and Engineering*, *390*(012075), 1–6. https://doi.org/10.1088/1757-899X/390/1/012075

Parswajinan, C., Ramnath, B. V., Hilary, I. A., Ram, N. C., & Mayandi, A. (2018b). Investigation of mechanical properties of aluminium reinforced Fly ash with CNT investigation of mechanical properties of aluminium reinforced Fly ash with CNT. *IOP Conf. Series: Materials Science and Engineering*, *390*(012032), 1–6. https://doi.org/10.1088/1757-899X/390/1/012032

Perez, N. (2004). *Electrochemistry and corrosion science*. Kluwer Academic Publishers.

Perrault, G. G. (1979). The role of hydrides in the equilibrium of aluminium in aqueous solutions. *Journal of the Electrochemical Society*, *126*(2), 199–204.

Petrović, J., Mladenović, S., Marković, I., & Dimitrijević, S. (2022). Characterization of hybrid Aluminium composites reinforced with Al_2O_3 particles and walnut-shell ash. *Materiali in Tehnologije*, *56*(2), 115–122. https://doi.org/10.17222/mit.2022.365

Poornesh, M., Harish, N., & Aithal, K. (2016). Study of mechanical properties of aluminium alloy composites. *American Journal of Materials Science*, *6*(4A), 72–76. https://doi.org/10.5923/c.materials.201601.14

Prasad, D. S., & Rama Krishna, A. (2012). Tribological properties of A356.2/RHA composites. *Journal of Materials Science and Technology*, *28*(4), 367–372.

Prasad, D. S., & Shoba, C. (2014). Hybrid composites- A better choice for high wear resistant. *Journal of Materials Research and Technology*, *3*(2), 172–178.

Prasad, D. V., Shoba, C., & Ramanaiah, N. (2014). Investigations of mechanical properties of Aluminium hybrid composites. *Journal of Materials Research and Technology*, *3*(1), 79–85. https://doi.org/10.1016/j.jmrt.2013.11.002

Prasad, N. (2006). Development and characterization of metal matrix composite using red mud an industrial waste for wear resistant applicants [Doctoral Dissertation, National Institute of Technology, Rourkela-769 008, India]. https://www.academia.edu/50583943/development_and_characterization_of_metal_matrix_composite_using_red_mud_an_industrial_waste_for_wear_resistance_applications

Prasad, S. V., & Asthana, R. (2004). Aluminium metal-matrix composites for automotive applications: Tribological considerations. *Tribology Letters*, *17*(3), 445–453. https://doi.org/10.1023/B:TRIL.0000044492.91991.f3

Raei, M., Panjepour, M., & Meratian, M. (2016). Effect of stirring speed and time on microstructure and mechanical properties of Cast Al – Ti – Zr – B 4 C composite produced by stir casting 1. *Russian Journal of Non-Ferrous Metals*, *57*(4), 347–360. https://doi.org/10.3103/S1067821216040088

Rahman, A. A., Sulaiman, F., & Abdullah, N. (2016). Influence of washing medium pre-treatment on pyrolysis yields and product characteristics of palm kernel shell. *Journal of Physical Science*, *27*(1), 53–75.

Rajesh, A. M., & Kaleemulla, M. (2016). Experimental investigations on mechanical behaviour of aluminium metal matrix composites. *IOP Conference Series: Materials Science and Engineering*, *149*(1). https://doi.org/10.1088/1757-899X/149/1/012121

Rajesh, A., & Santosh, D. (2017). Mechanical properties of Al-SiC metal matrix composites fabricated by stir casting route. *Research Medical Engineering Science*, *2*(6), 1–6.

Ramnath, B. V., Elanchezhian, C., Jaivignesh, M., Rajesh, S., Parswajinan, C., & Siddique, A. G. A. (2014). Evaluation of mechanical properties of aluminium alloy–alumina–boron carbide metal matrix composites. *Materials and Design*, *58*, 332–338.

Ratna Kumar, P. S. S., Robinson Smart, D. S., & Alexis, S. J. (2017). Corrosion behaviour of aluminium metal matrix reinforced with multi-wall carbon nanotube. *Journal of Asian Ceramic Societies*, *5*(1), 71–75. https://doi.org/10.1016/j.jascer.2017.01.004

Ravindran, P., Manisekar, K., Vinoth Kumar, S., & Rathika, P. (2013). Investigation of microstructure and mechanical properties of Aluminium hybrid nano-composites with the additions of solid lubricant. *Materials and Design*, *51*(448–456). https://doi.org/10.1016/j.matdes.2013.04.015

Reddy, B. R., & Srinivas, C. (2018). Fabrication and characterization of silicon carbide and fly ash reinforced aluminium metal matrix hybrid composites. *Materials Today: Proceedings*, *5*(2). https://doi.org/10.1016/j.matpr.2017.11.531

Sahu, M. K., & Sahu, R. K. (2018). Fabrication of of Aluminium Aluminium Matrix Matrix Composites Composites by Stir Stir Casting Technique Technique and and Stirring Stirring Process Process Parameters Parameters Optimization. In *Advanced Casting Technologies* (Issue May, pp. 111–126). Intech Open. https://doi.org/10.5772/intechopen.73485

Sambathkumar, M., Sasikumar, K. S. K., Gukendran, R., Dineshkumar, K., Ponappa, K., & Harichandran, S. (2021). Investigation of mechanical and corrosion properties of Al 7075/Redmud metal matrix composite. *Revista de Metalurgia*, *57*(1), 1–9. https://doi.org/10.3989/REVMETALM.185

Senapati, A. K., & Mohanta, G. K. (2016). Experimental study on mechanical properties of aluminium alloy reinforced with silicon carbide and fly ash, hybrid metal matrix composites. *International Journal of Advanced Research in Science and Engineering*, *5*(1), 457–463.

Seyed Reihani, S. M. (2006). Processing of squeeze cast Al6061–30 vol% SiC composites and their characterization. *Materials and Design*, *27*, 216–222.

Singh, G., & Goyal, S. (2018). Microstructure and mechanical behaviour of AA6082-T6/SiC/B4C-based Aluminium hybrid composites. *Particulate Science and Technology*, *36*(2), 154–161. https://doi.org/10.1080/02726351.2016.1227410

Smith, W. F., & Hashemi, J. (2008). *Materials science and engineering*. Tata McGraw Hill Education Private Limited.

Sree Manu, K.M., Ajay Raag, L., Rajan, T.P.D., Gupta, M., & Pai, B.C. (2016). Liquid metal infiltration processing of metallic composites: A critical review. *Metallurgical and Materials Transactions B*, 47, 2799–2819.

Sukiman, N. L., Zhou, X., Birbilis, N., Hughes, A. E., Mol, J. M. C., Garcia, S. J., Zhou, X., & Thompson, G. E. (2012). Durability and corrosion of aluminium and its alloys: Overview, property space, techniques and developments. In Z. Ahmad (Ed.), *Aluminium alloys new trends in fabrication and applications* (pp. 224–262). Intechopen. https://doi.org/10.5772/53752

Sukumaran, K., Ravikumar, K. K., Pillai, S. G. K., Rajan, T. P. D., Ravi, M., Pillai, R. M., & Pai, B. C. (2008). Studies on squeeze casting of Al2124 alloy and 2124-10%SiCp metal matrix composite. *Materials Science and Engineering A*, 235–241.

Surappa, M. K. (2003). Aluminium matrix composites: Challenges and opportunities. *Sadhana*, *28*(1&2), 319–334.

Suresh Kumar, S., Thirumalai Kumaran, S., Velmurugan, G., Perumal, A., Sekar, S., & Uthayakumar, M. (2022). Physical and mechanical properties of various metal matrix composites: A review. *Materials Today: Proceedings*, *50*(5), 1022–1031. https://doi.org/10.1016/j.matpr.2021.07.354

Surya, M. S., & Gugulothu, S. K. (2022). Fabrication, mechanical and wear characterization of silicon carbide reinforced aluminium 7075 metal matrix composite. *Silicon*, *14*(5), 2023–2032. https://doi.org/10.1007/s12633-021-00992-x

Tangchirapat, W., Saeting, T., Jaturapitakkul, C., Kiattikomol, K., & Siripanichgorn, A. (2007). Use of waste ash from palm oil industry in concrete. *Waste Management*, *27*(1), 81–88. https://doi.org/10.1016/j.wasman.2005.12.014

Wang, X., Chen, G., Yang, W., Wu, G., & Jiang, D. (2010). Effect of magnesium content on the corosion behaviors of Gr_f/Al composite. *Metallurgical and Materials Transactions A: Physical Metallurgy and Materials Science*, *4*(13), 3458–3462. https://doi.org/10.1007/s11661-010-0509-7

Xavier, L. F., & Suresh, P. (2016). Wear behaviour of aluminium metal matrix composite prepared from industrial waste. *Scientific World Journal*, *2016*, 1–8. https://doi.org/10.1155/2016/6538345

Yacob, A. R., & Noramirah, W., Nurshaira, H.S., and Mohd, K. A. A. (2013). Microwave induced carbon from waste palm kernel shell activated by phosphoric acid. *International Journal of Engineering and Technology*, *5*(2), 214–217. https://doi.org/10.7763/ijet.2013.v5.545

Yin, C., Rosendahl, L. A., & Kær, S. K. (2008). Grate-firing of biomass for heat and power production. *Progress in Energy and Combustion Science*, *34*(6), 725–754. https://doi.org/10.1016/j.pecs.2008.05.002

Yue, T. M., & Chadwick, G. A. (1996). Squeeze casting of light alloys and their composites. *Journal of Materials Processing Technology*, *58*, 302–307.

Yusuf, I. T., & Jimoh, Y. A. (2011). Palm kernel shell waste recycled concrete road as cheap and environmental friendly freeway on very poor subgrades. *National Engineering Conference (Nigerian Society of Engineers)*, 1–7.

Zakaria, H. M. (2014). Microstructural and corrosion behaviour of Al/SiC metal matrix composites. *Ain Shams Ehgineering Journal*, *5*, 831–838. https://doi.org/10.1016/j.asej.2014.03.003

Zemlianov, A., Balokhonov, R., Romanova, V., & Gaiyatullina, D. (2022). The influence of bi-layer metal-matrix composite coating on the strength of the coated material. *Procedia Structural Intergrity*, *35*, 181–187. https://doi.org/10.1016/j.prostr.2021.12.063

12 Sustainable Innovations in Nanofiber Fabrication
Exploring the Latest Trends

Sagarika Bhattacharjee and Harmanpreet Singh
Indian Institute of Technology Ropar, Rupnagar, India

A. A. Adeleke
Nile University of Nigeria, Abuja, Nigeria

P. P. Ikubanni
Landmark University, Omu-Aran, Nigeria

12.1 INTRODUCTION

As we know, day by day, a lot of technological reforms are taking place. So, researchers are interested in exploring more materials and combining various properties into one single material, such as metal composites, smart materials, etc. Recently, a lot of work has been done in the field of nanotechnology. Nanotechnology is the vast field that deals with the designs, followed by production, and also the application of various structures, devices, and systems in the nanoscale i.e., dimensions in the range of 1–100 nm. Due to the larger surface area of such smaller particles, we find many distinctive properties of materials that are usually not found in bulk nature. As all new technology faces issues during implementation, so too did nanomaterials take their time to be introduced in the global market. There are many such companies and laboratories across the globe where large-scale production of nanomaterials has started to facilitate their multifunctionality in various sectors like petroleum, paints and dyes, construction components, electronics and communication, lubrication applications, insulating materials, bioengineering, drugs, medicine, and many more.

DOI: 10.1201/9781003309123-12

12.2 BIOPRINTING

Bioprinting or 3D bioprinting can be defined as a type of additive manufacturing process in which bioinks are used as a feed material to be printed layer by layer. The advanced cellular scaffolds, tissue constructs, and in vitro models are automatically fabricated through 3D cellular constructs (Di Marzio et al., 2020). This technology has opened up large opportunities in the advancement of bioengineering and the biomedical field. For example, If a patient has lost any limb or a major part of his organs, then using 3D bioprinting new limbs or the organ part can be designed similarly to the original one within a few hours. So, there won't be any difficulty with blood transfusion or matching of the organ donors, etc.

Scaffolds are responsible for providing the proper micro-environment for cell growth along with structural support. For successful tissue repair, the scaffold should imitate the normal tissue in the context of organized structure as well as in its physiological functioning. There can be direct maturation of the bio-printed model into the functional tissue where the cell can be seeded or grown. The three major approaches for scaffold synthesis include electrospinning (Buttafoco et al., 2006; Chen et al., 2015; Sell et al., 2009), phase separation (Heijkants et al., 2008; Nieponice et al., 2010; Y. C. Wang et al., 2003), and self-assembly (P. X. Ma & Zhang, 1999). Such types of methods are discussed next.

12.3 NANOFIBER

Nanofibers are simply the fibers with their characteristic dimension in nanoscale, i.e., 1–100 nm. Nanofibers are used in multiple biomedical domains, including drug delivery, healing of wounds, tissue engineering, coatings of the implants, etc. There large ratio of surface-to-volume helps in increasing the activity of the materials.

12.3.1 Properties

It is found that the constructs with embedded nanofibers have a fourfold enhancement in the mechanical properties, including resilience limit, compressive strength, and modulus (Yoon et al., 2019). Thus, when nanofibers are compared with macrofibres, some of the exclusive properties are never obtained in macrofibres. For example, due to the presence of high surface area along with the different means of composition adjustment through the fabrication process, nanofibers accelerate wound healing (Leung & Ko, 2011).

12.3.2 Fabrication Methods

12.3.2.1 Electrospinning

Electrospinning is a very well-known fabrication technique for nanofibers that is frequently used in various applications. In this process, there is the use of electric force to obtain the fibers, particularly with a diameter of order nanometers or a few microns from the charged threads of polymer solutions/melts. There are some common reasons why electrospinning is used so widely. They are as follows:

- Low-cost method
- Easy to control fiber morphological parameters like thickness, length, etc.
- The capability to scale up for mass production
- It can be tailored to suit different types of tissues or to load various drugs (Al-Hazeem, 2018).

Electrospinning has three basic main components. They are

- a power supply with high voltage,
- a reservoir with a polymer solution (which includes the thin diameter needle at the end), and
- a metal collecting screen (Figure 12.1).

The controlled and adjustable high voltage is supplied to obtain the approximate DC output of about 50 kV, wherein the multiple outputs are achieved which depends upon the number of electrospinning jets. The reservoir with the polymeric solution is connected to a power supply to obtain the charged polymer jet.

When the potential is increased, a cone-like structure is obtained as the hemispherical surface of the solution is elongated at the tip of the capillary tube, and the

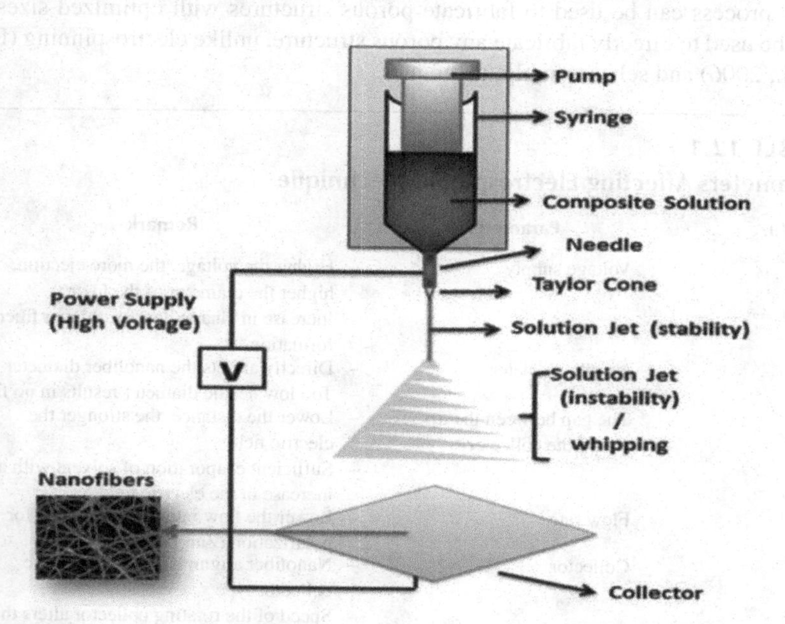

FIGURE 12.1 Schematic diagram of electrospinning setup (Al-Hazeem, 2018).

Source: Al-Hazeem, N. Z. (2018). Nanofibers and electrospinning method. *Nanomaterials – Synthesis and Application*. Used under CC BY-3.0.

cone thus obtained is known as the Taylor cone (TC). The surface tension forces overcome to form a jet with the achievement of the critical limit in electric potential, and the jet is ejected from the tip of the TC. Further, the charged jet gets converted to the randomly oriented nanofibers collected by the rotating metallic collector.

Electrospun nanofibers are very useful for the fabrication of tissue scaffolds due to their versatility as well as the tailorable property of the electrospinning process for specific tissue applications (Table 12.1).

12.3.2.2 Freeze-Drying

Freeze-drying, as the name suggests, involves very low-temperature application. This process is also called ice segregation-induced self-assembly. This technique is majorly divided into three steps. They are as follows:

1. A solution is prepared with the required precursor, which is frozen at low temperatures, i.e. (−20 to −100°C), allowing ice crystal growth and nucleation.
2. Primary drying process by removing the ice from the frozen sample using sublimation. Sublimation is achieved by reducing the pressure with the help of a vacuum.
3. The rest of the unfrozen water left in the system is removed using desorption as a secondary drying process.

This process can be used to fabricate porous structures with optimized sizes. It can be used to directly fabricate any porous structure, unlike electrospinning (Rho et al., 2006) and self-assembly techniques.

TABLE 12.1
Parameters Affecting Electrospinning Technique

Sr. No.	Parameter	Remark
1	Voltage supply	– Higher the voltage, the more ejection, higher the diameter of the fiber – Increase in charge density, thinner fiber formation
2	Needle diameter	– Directly affects the nanofiber diameter – Too low needle diameter results in no flow
3	The gap between the tip and the collector	– Lower the distance, the stronger the electric field – Sufficient evaporation of solvent with an increase in the electric field
4	Flow rate	– Lower the flow rate, more the time for polarization (Zargham et al., 2012)
5	Collector	– Nanofiber alignment depends on the collector type – Speed of the rotating collector alters the alignment of nanofibers

A lot of experiments have been carried out using chitin. J. Carson Meredith et al. have demonstrated that by adjusting the variables it is possible to tune the dimensions and networking of the fiber formation (J. Wu & Meredith, 2014). It was observed that with a change in the freezing temperature, the structures formed were also changed. For example: At −80°C, the frozen chitin consists of sheet-like random porous structures. On the other hand, Chitosan forms microporous sheet-like structures at −20°C. It is believed that the solubility of chitosan leads to precipitation into larger domains during the time of freezing. At the same time, insoluble chitin nanofibers lead to the formation of uniform nanostructures (J. Wu & Meredith, 2014).

12.3.2.3 Template Synthesis

The process of template synthesis fabricates the solid nanofibers in existence into nanoporous membrane template (Yan et al., 2019). Fabrication of nanofibers with a variety of materials viz metals, conductive polymers, and semiconductors can be achieved through this process. The pore size of nanofibers can be controlled using a membrane in this process. Although it is a popular technique, it is not quite often used in the fabrication of nanofibers. This is because we usually end up getting mesoporous structures instead of nanoporous. For example, C. G. Wu et al. synthesize conducting polyaniline filaments in a mesoporous channel host using the template synthesis method (C. G. Wu & Bein, 1994). He prepared the conducting filaments in the 3 nm wide hexagonal channel system of the aluminosilicate MCM-4 (C. G. Wu & Bein, 1994).

12.3.2.4 Thermal-Induced Phase Separation (TIPS)

The scaffolds with interconnected porous structures can be obtained mostly by following the procedure of TIPS. The multiphase system is created by the demixing of the homogeneous polymeric solution, which is achieved by the temperature variation.

The general procedure of nanofiber fabrication using TIPS consists of the following steps (Lloyd et al., 1990):

- A homogeneous melt blend is formed by dissolving the desired polymer in a solvent with a high boiling point with low molecular weight at an elevated temperature near or higher than the melting point of the polymer.
- Then, the dope solution is given the desired shape, like a hollow fiber or flat sheet.
- The casted solution is then cooled in a controlled way so that phase separation and precipitation are induced in the polymer (Figure 12.2).

The TIPS process is a well-known method for obtaining membranes with controlled morphology (Ji et al., 2007; Mulder, 1996). Shang et al. used the mixture

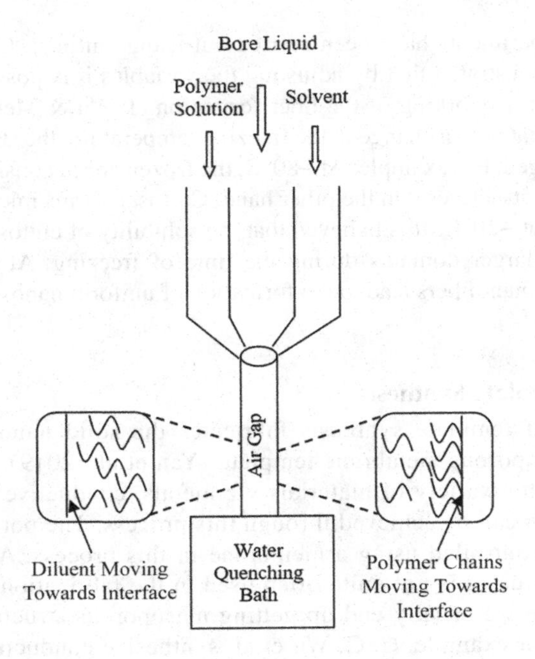

FIGURE 12.2 Setup of TIPS.

TABLE 12.2
Parameters Affecting TIPS

Sr. No.	Parameter	Remark
1	Polymer Concentration	– The higher the concentration, the lower the pore dimension – Porosity is inversely proportional to the polymer concentration
2	Solvent/non-solvent ratio	– The progress in the crystallization of the solvent dominates the scaffold structure – Adding a non-solvent improves the scaffold architecture with increased sensitivity
3	Cooling path	– the pore dimension multiplies itself with an increase in time (Mannella et al., 2014, 2015)

of 1,3-propanediol and glycerol as diluents in the synthesis of poly(ethylene-co-vinyl alcohol) hollow-fiber membranes via TIPS method (Shang et al., 2005). To enhance the mechanical properties and bioactivity of the scaffold, an inorganic filler can be incorporated suitably into the scaffold matrix in a single step. Table 12.2, gives a brief about the operating parameters.

12.4 APPLICATIONS

In the field of bioengineering and biomedicine, nanofibers have tremendous applications. To be precise, Tissue Engineering and Regenerative Medicine (TERM) and drug delivery systems got a thrust for enhancement of technologies using the above-discussed nanofiber fabrication techniques. When TERM is concerned, the property of crosslinking of fibers plays a vital role. Hence, various process parameters, as discussed earlier, help to meet the requirement.

With structural and mechanical supports, it was observed that perfect perfusion condition for the growth of cells is provided by the 3D fabricated complex constructs, which also include the fully opened internal channels (Yoon et al., 2019). One of the very important applications of nanofiber is wound healing. Nanofiber of decomposable polymer creates the fibrous mat dressing when it comes in direct contact with the wounded skin. This mechanism enhances the skin formation on the wound as well as minimizes the formation of wound tissues (J. Wu & Meredith, 2014) (Table 12.3).

TABLE 12.3
Summary of Fabrication Techniques Used for Biomaterials or Bio-Composites for Various Applications

Sr. No.	Biomaterials/ Bio-Composites	The Method of Fabrication	Application	Reference
1	Chitosan	Freeze-drying	Scaffold	(Nwe et al., 2009)
2	Collagen	Freeze-drying	Scaffold	(Oh et al., 2012)
3	Collagen	Freeze-drying	Drug delivery system	(Nanda et al., 2014)
4	PLGA	Thermal-induced phase separation	Scaffold	(Day et al., 2004)
5	Chitosan/PLA	Freeze-drying	Scaffold	(Zhang & Cui, 2012)
6	PLA/ nanocellulose	Electrospinning	Release of non-ionic compounds	(Xiang et al., 2013)
7	Cellulose/ chitosan	Freeze-drying	Sorption of trimethylamine and metal ions	(Twu et al., 2003)
8	Cellulose/ chitosan	Freeze-drying	Dye adsorption	(Li et al., 2015)
9	Bacterial cellulose nanofiber/ chitosan	Freeze-drying	Scaffold	(J. Kim et al., 2011)
10	Chitosan/PLGA	Electrospinning	Scaffold	(S. J. Kim et al., 2013)
11	Chitosan/PLGA nanocomposite	Electrospinning and freeze-drying	Scaffold	(Cui et al., 2014)
12	Chitosan/PLGA nanocomposite	Electrospinning and unidirectional freeze-drying	Scaffold	(Yuanyuan & Song, 2012)

(Continued)

TABLE 12.3
(Continued)

Sr. No.	Biomaterials/ Bio-Composites	The Method of Fabrication	Application	Reference
13	Chitosan/collagen	Freeze-drying	Scaffold	(L. Ma et al., 2003; Peng et al., 2006; L. Wang & Stegemann, 2011)
14	Poly (vinylidene fluoride)	TIPS	Membranes	(Ji et al., 2007)
15	Poly (ethylene-co-vinyl alcohol)	TIPS	Hollow-fiber membranes	(Shang et al., 2005)
16	Polypropylene	TIPS	Microporous membrane	(Lloyd et al., 1990)
17	Polyaniline	Template synthesis	Filaments	(C. G. Wu & Bein, 1994)

12.5 DRAWBACKS

In the drug delivery system and the tissue engineering field, it is clear that there are various other methods of nanofiber fabrication other than electrospinning. But certain aspects make it less viable to be used.

First, not all polymers or bioinks can be used for nanofiber fabrication through recent techniques such as freeze-drying or TIPS. Most of the bioinks tend to form direct sheets or use polymer solutions as a precursor to obtaining various scaffolds using 3D printers. Comparatively, electrospinning can be widely used, provided that the precursor has charged particles.

Second, the strength of the nanofiber formed after fabrication depends largely on the type of fabrication process. For example, in the case of freeze-drying, the strength of the nanofiber synthesized would largely depend upon the additives or polymers included in the melt or solution before processing. Without the use of any of the additives, the nanofibers formed would be very fragile and easily deformed (depending on the working temperature).

Another aspect that can be considered as a disadvantage is that most of these nanofiber fabrication processes take a bit more time than expected. To overcome the time constraint, other 3D printing methods are followed that do not require nanofiber fabrication in the first place, such as laser-assisted bioprinting (Loai et al., 2019), stereolithography (Malda et al., 2013; Z. Wang et al., 2018), etc.

12.6 SUMMARY

In the field of bioengineering and biomedical applications, nanofibers have been contributing largely to the sector. Although electrospinning is considered to be the most commonly used method of fabrication of nanofiber, there is still a need for other methods too. In view of recent trends, we conclude that electrospinning is still the evergreen technique to be used for nanofiber production. Other than that, freeze-drying and TIPS have their specialties. Also, it was observed that to shorten the total time of processing, other 3D printing methods that help in omitting a few processing steps are adopted these days.

REFERENCES

Al-Hazeem, N. Z. A. (2018). *Nanofibers and Electrospinning Method* (G. Z. Kyzas & A. C. Mitropoulos (eds.); p. Ch. 11). IntechOpen. https://doi.org/10.5772/intechopen.72060

Buttafoco, L., Kolkman, N. G., Engbers-Buijtenhuijs, P., Poot, A. A., Dijkstra, P. J., Vermes, I., & Feijen, J. (2006). Electrospinning of collagen and elastin for tissue engineering applications. *Biomaterials*, *27*(5), 724–734. https://doi.org/10.1016/j.biomaterials.2005.06.024

Chen, W., Li, D., Ei-Shanshory, A., El-Newehy, M., Ei-Hamshary, H. A., Al-Deyab, S. S., He, C., & Mo, X. (2015). Dexamethasone loaded core-shell SF/PEO nanofibers via green electrospinning reduced endothelial cells inflammatory damage. *Colloids and Surfaces. B, Biointerfaces*, *126*, 561–568. https://doi.org/10.1016/j.colsurfb.2014.09.016

Cui, Z., Zhao, H., Peng, Y., Han, J., Turng, L.-S., & Shen, C. (2014). Fabrication and Characterization of Highly Porous Chitosan/Poly(DL lactic-co-glycolic acid) Nanocomposite Scaffolds Using Electrospinning and Freeze Drying. *Journal of Biobased Materials and Bioenergy*, *8*(3), 281–291. https://doi.org/10.1166/jbmb.2014.1444

Day, R. M., Boccaccini, A. R., Maquet, V., Shurey, S., Forbes, A., Gabe, S. M., & Jérôme, R. (2004). In vivo characterisation of a novel bioresorbable poly(lactide-co-glycolide) tubular foam scaffold for tissue engineering applications. *Journal of Materials Science: Materials in Medicine*, *15*(6), 729–734. https://doi.org/10.1023/B:JMSM.0000030216.73274.86

Di Marzio, N., Eglin, D., Serra, T., & Moroni, L. (2020). Bio-Fabrication: Convergence of 3D Bioprinting and Nano-Biomaterials in Tissue Engineering and Regenerative Medicine. *Frontiers in Bioengineering and Biotechnology*, *8*(April). https://doi.org/10.3389/fbioe.2020.00326

Heijkants, R. G. J. C., van Calck, R. V., van Tienen, T. G., de Groot, J. H., Pennings, A. J., Buma, P., Veth, R. P. H., & Schouten, A. J. (2008). Polyurethane scaffold formation via a combination of salt leaching and thermally induced phase separation. *Journal of Biomedical Materials Research. Part A*, *87*(4), 921–932. https://doi.org/10.1002/jbm.a.31829

Ji, G., Du, C., Zhu, B., & Xu, Y. (2007). Preparation of porous PVDF membrane via thermally induced phase separation with diluent mixture of DBP and DEHP. *Journal of Applied Polymer Science*, *105*(3), 1496–1502. https://doi.org/10.1002/app.26385

Kim, J., Cai, Z., Lee, H. S., Choi, G. S., Lee, D. H., & Jo, C. (2011). Preparation and characterization of a Bacterial cellulose/Chitosan composite for potential biomedical application. *Journal of Polymer Research*, *18*(4), 739–744. https://doi.org/10.1007/s10965-010-9470-9

Kim, S. J., Yang, D. H., Chun, H. J., Chae, G. T., Jang, J. W., & Shim, Y. B. (2013). Evaluations of chitosan/poly(D,L-lactic-co-glycolic acid) composite fibrous scaffold for tissue engineering applications. *Macromolecular Research*, *21*(8), 931–939. https://doi.org/10.1007/s13233-013-1110-x

Leung, V., & Ko, F. (2011). Biomedical applications of nanofibers. *Polymers for Advanced Technologies*, *22*(3), 350–365. https://doi.org/10.1002/pat.1813

Li, M., Wang, Z., & Li, B. (2015). Adsorption behaviour of congo red by cellulose/chitosan hydrogel beads regenerated from ionic liquid. *Desalination and Water Treatment*, 1–11. https://doi.org/10.1080/19443994.2015.1082945

Lloyd, D. R., Kinzer, K. E., & Tseng, H. S. (1990). Microporous membrane formation via thermally induced phase separation. I. Solid-liquid phase separation. *Journal of Membrane Science*, *52*(3), 239–261. https://doi.org/10.1016/S0376-7388(00)85130-3

Loai, S., Kingston, B., Wang, Z., Philpott, D., Tao, M., & Cheng, H. (2019). Clinical Perspectives on 3D Bioprinting Paradigms for Regenerative Medicine. *Regenerative Medicine Frontiers*. https://doi.org/10.20900/rmf20190004

Ma, L., Gao, C., Mao, Z., Zhou, J., Shen, J., Hu, X., & Han, C. (2003). Collagen/chitosan porous scaffolds with improved biostability for skin tissue engineering. *Biomaterials*, *24*(26), 4833–4841. https://doi.org/10.1016/S0142-9612(03)00374-0

Ma, P. X., & Zhang, R. (1999). Synthetic nano-scale fibrous extracellular matrix. *Journal of Biomedical Materials Research*, *46*(1), 60–72. https://doi.org/10.1002/(SICI)1097-4636(199907)46:1<60::AID-JBM7>3.0.CO;2-H

Malda, J., Visser, J., Melchels, F. P., Jüngst, T., Hennink, W. E., Dhert, W. J. A., Groll, J., & Hutmacher, D. W. (2013). 25th Anniversary Article: Engineering Hydrogels for Biofabrication. *Advanced Materials*, *25*(36), 5011–5028. https://doi.org/10.1002/adma.201302042

Mannella, G. A., Carfì Pavia, F., Conoscenti, G., La Carrubba, V., & Brucato, V. (2014). Evidence of mechanisms occurring in thermally induced phase separation of polymeric systems. *Journal of Polymer Science Part B: Polymer Physics*, *52*(14), 979–983. https://doi.org/10.1002/polb.23518

Mannella, G. A., Conoscenti, G., Carfì Pavia, F., La Carrubba, V., & Brucato, V. (2015). Preparation of polymeric foams with a pore size gradient via Thermally Induced Phase Separation (TIPS). *Materials Letters*, *160*, 31–33. https://doi.org/10.1016/j.matlet.2015.07.055

Mulder, M. (1996). *Basic Principles of Membrane Technology*. Springer Netherlands. https://doi.org/10.1007/978-94-009-1766-8

Nanda, H. S., Chen, S., Zhang, Q., Kawazoe, N., & Chen, G. (2014). Collagen Scaffolds with Controlled Insulin Release and Controlled Pore Structure for Cartilage Tissue Engineering. *BioMed Research International*, *2014*, 1–10. https://doi.org/10.1155/2014/623805

Nieponice, A., Soletti, L., Guan, J., Hong, Y., Gharaibeh, B., Maul, T. M., Huard, J., Wagner, W. R., & Vorp, D. A. (2010). In vivo assessment of a tissue-engineered vascular graft combining a biodegradable elastomeric scaffold and muscle-derived stem cells in a rat model. *Tissue Engineering. Part A*, *16*(4), 1215–1223. https://doi.org/10.1089/ten.TEA.2009.0427

Nwe, N., Furuike, T., & Tamura, H. (2009). The mechanical and biological properties of Chitosan Scaffolds for tissue regeneration templates are significantly enhanced by Chitosan from Gongronella butleri. *Materials*, *2*(2), 374–398. https://doi.org/10.3390/ma2020374

Oh, H. H., Ko, Y., Lu, H., Kawazoe, N., & Chen, G. (2012). Preparation of porous collagen scaffolds with micropatterned structures. *Advanced Materials*, *24*(31), 4311–4316. https://doi.org/10.1002/adma.201200237

Peng, L., Cheng, X. R., Wang, J. W., Xu, D. X., & Wang, G. (2006). Preparation and evaluation of porous Chitosan/Collagen scaffolds for periodontal tissue engineering. *Journal of Bioactive and Compatible Polymers*, *21*(3), 207–220. https://doi.org/10.1177/0883911506065100

Rho, K. S., Jeong, L., Lee, G., Seo, B.-M., Park, Y. J., Hong, S.-D., Roh, S., Cho, J. J., Park, W. H., & Min, B.-M. (2006). Electrospinning of collagen nanofibers: Effects on the behavior of normal human keratinocytes and early-stage wound healing. *Biomaterials*, *27*(8), 1452–1461. https://doi.org/10.1016/j.biomaterials.2005.08.004

Sell, S. A., McClure, M. J., Garg, K., Wolfe, P. S., & Bowlin, G. L. (2009). Electrospinning of collagen/biopolymers for regenerative medicine and cardiovascular tissue engineering. *Advanced Drug Delivery Reviews*, *61*(12), 1007–1019. https://doi.org/10.1016/j.addr.2009.07.012

Shang, M., Matsuyama, H., Teramoto, M., Okuno, J., Lloyd, D. R., & Kubota, N. (2005). Effect of diluent on poly(ethylene- co -vinyl alcohol) hollow-fiber membrane formation via thermally induced phase separation. *Journal of Applied Polymer Science*, *95*(2), 219–225. https://doi.org/10.1002/app.21193

Twu, Y.-K., Huang, H.-I., Chang, S.-Y., & Wang, S.-L. (2003). Preparation and sorption activity of chitosan/cellulose blend beads. *Carbohydrate Polymers*, *54*(4), 425–430. https://doi.org/10.1016/j.carbpol.2003.03.001

Wang, L., & Stegemann, J. P. (2011). Glyoxal crosslinking of cell-seeded chitosan/collagen hydrogels for bone regeneration. *Acta Biomaterialia*, *7*(6), 2410–2417. https://doi.org/10.1016/j.actbio.2011.02.029

Wang, Y. C., Lin, M. C., Wang, D. M., & Hsieh, H. J. (2003). Fabrication of a novel porous PGA-chitosan hybrid matrix for tissue engineering. *Biomaterials*, *24*(6), 1047–1057. https://doi.org/10.1016/s0142-9612(02)00434-9

Wang, Z., Kumar, H., Tian, Z., Jin, X., Holzman, J. F., Menard, F., & Kim, K. (2018). Visible Light Photoinitiation of Cell-Adhesive Gelatin Methacryloyl Hydrogels for Stereolithography 3D Bioprinting. *ACS Applied Materials & Interfaces*, *10*(32), 26859–26869. https://doi.org/10.1021/acsami.8b06607

Wu, C. G., & Bein, T. (1994). Conducting polyaniline filaments in a mesoporous channel host. *Science (New York, N.Y.)*, *264*(5166), 1757–1759. https://doi.org/10.1126/science.264.5166.1757

Wu, J., & Meredith, J. C. (2014). Assembly of Chitin nanofibers into porous biomimetic structures via freeze drying. *ACS Macro Letters*, *3*(2), 185–190. https://doi.org/10.1021/mz400543f

Xiang, C., Taylor, A. G., Hinestroza, J. P., & Frey, M. W. (2013). Controlled release of nonionic compounds from poly(lactic acid)/cellulose nanocrystal nanocomposite fibers. *Journal of Applied Polymer Science*, *127*(1), 79–86. https://doi.org/10.1002/app.36943

Yan, J., Han, Y., Xia, S., Wang, X., Zhang, Y., Yu, J., & Ding, B. (2019). Polymer template synthesis of flexible BaTiO3 crystal nanofibers. *Advanced Functional Materials*, *29*(51), 1907919. https://doi.org/10.1002/adfm.201907919

Yoon, Y., Kim, C. H., Lee, J. E., Yoon, J., Lee, N. K., Kim, T. H., & Park, S.-H. (2019). 3D bioprinted complex constructs reinforced by hybrid multilayers of electrospun nanofiber sheets. *Biofabrication*, *11*(2), 25015. https://doi.org/10.1088/1758-5090/ab08c2

Yuanyuan, Z., & Song, L. (2012). Preparation of chitosan/poly (lactic-co glycolic acid) (PLGA) nanocoposite for tissue engineering scaffold. *Optoelectronics and Advanced Materials, Rapid Communications*, *6*, 516–519.

Zargham, S., Bazgir, S., Tavakoli, A., Rashidi, A. S., & Damerchely, R. (2012). The effect of flow rate on morphology and deposition area of electrospun nylon 6 nanofiber. *Journal of Engineered Fibers and Fabrics*, *7*(4), 42–49. https://doi.org/10.1177/155892501200700414

Zhang, Z., & Cui, H. (2012). Biodegradability and biocompatibility study of poly(Chitosan-g-lactic Acid) scaffolds. *Molecules*, *17*(3), 3243–3258. https://doi.org/10.3390/molecules17033243

Index

Pages in *italics* refer to figures and pages in **bold** refer to tables.

D

E

F

G

H

Printed in the United States
by Baker & Taylor Publisher Services

Printed in the United States
by Baker & Taylor Publisher Services